Selected Titles in This Series

277 **Luca Capogna and Loredana Lanzani, Editors,** Harmonic analysis and boundary value problems, 2001

276 **Emma Previato, Editor,** Advances in algebraic geometry motivated by physics, 2001

275 **Alfred G. Noël, Earl Barnes, and Sonya A. F. Stephens, Editors,** Council for African American researchers in the mathematical sciences: Volume III, 2001

274 **Ken-ichi Maruyama and John W. Rutter, Editors,** Groups of homotopy self-equivalences and related topics, 2001

273 **A. V. Kelarev, R. Göbel, K. M. Rangaswamy, P. Schultz, and C. Vinsonhaler, Editors,** Abelian groups, rings and modules, 2001

272 **Eva Bayer-Fluckiger, David Lewis, and Andrew Ranicki, Editors,** Quadratic forms and their applications, 2000

271 **J. P. C. Greenlees, Robert R. Bruner, and Nicholas Kuhn, Editors,** Homotopy methods in algebraic topology, 2001

270 **Jan Denef, Leonard Lipschitz, Thanases Pheidas, and Jan Van Geel, Editors,** Hilbert's tenth problem: Relations with arithmetic and algebraic geometry, 2000

269 **Mikhail Lyubich, John W. Milnor, and Yair N. Minsky, Editors,** Laminations and foliations in dynamics, geometry and topology, 2001

268 **Robert Gulliver, Walter Littman, and Roberto Triggiani, Editors,** Differential geometric methods in the control of partial differential equations, 2000

267 **Nicolás Andruskiewitsch, Walter Ricardo Ferrer Santos, and Hans-Jürgen Schneider, Editors,** New trends in Hopf algebra theory, 2000

266 **Caroline Grant Melles and Ruth I. Michler, Editors,** Singularities in algebraic and analytic geometry, 2000

265 **Dominique Arlettaz and Kathryn Hess, Editors,** Une dégustation topologique: Homotopy theory in the Swiss Alps, 2000

264 **Kai Yuen Chan, Alexander A. Mikhalev, Man-Keung Siu, Jie-Tai Yu, and Efim I. Zelmanov, Editors,** Combinatorial and computational algebra, 2000

263 **Yan Guo, Editor,** Nonlinear wave equations, 2000

262 **Paul Igodt, Herbert Abels, Yves Félix, and Fritz Grunewald, Editors,** Crystallographic groups and their generalizations, 2000

261 **Gregory Budzban, Philip Feinsilver, and Arun Mukherjea, Editors,** Probability on algebraic structures, 2000

260 **Salvador Pérez-Esteva and Carlos Villegas-Blas, Editors,** First summer school in analysis and mathematical physics: Quantization, the Segal-Bargmann transform and semiclassical analysis, 2000

259 **D. V. Huynh, S. K. Jain, and S. R. López-Permouth, Editors,** Algebra and its applications, 2000

258 **Karsten Grove, Ib Henning Madsen, and Erik Kjær Pedersen, Editors,** Geometry and topology: Aarhus, 2000

257 **Peter A. Cholak, Steffen Lempp, Manuel Lerman, and Richard A. Shore, Editors,** Computability theory and its applications: Current trends and open problems, 2000

256 **Irwin Kra and Bernard Maskit, Editors,** In the tradition of Ahlfors and Bers: Proceedings of the first Ahlfors-Bers colloquium, 2000

255 **Jerry Bona, Katarzyna Saxton, and Ralph Saxton, Editors,** Nonlinear PDE's, dynamics and continuum physics, 2000

For a complete list of titles in this series, visit the
AMS Bookstore at **www.ams.org/bookstore/**.

Harmonic Analysis and Boundary Value Problems

CONTEMPORARY MATHEMATICS

277

Harmonic Analysis and Boundary Value Problems

Selected Papers from the
25th University of Arkansas Spring Lecture Series
"Recent Progress in the Study of Harmonic Measure
from a Geometric and Analytic Point of View"
March 2–4, 2000
Fayetteville, Arkansas

Luca Capogna
Loredana Lanzani
Editors

American Mathematical Society
Providence, Rhode Island

This volume contains the proceedings of the 25th University of Arkansas Spring Lecture Series, "Recent Progress in the Study of Harmonic Measure from a Geometric and Analytic Point of View", held at the University of Arkansas, Fayetteville, March 2–4, 2000.

2000 *Mathematics Subject Classification.* Primary 35J25, 35J10, 35J67, 35J85, 31B20, 31B25, 31B35, 31A15.

Library of Congress Cataloging-in-Publication Data

Harmonic analysis and boundary value problems : selected papers from the 25th University of Arkansas spring lecture series, Recent progress in the study of harmonic measure from a geometric and analytic point of view, March 2–4, 2000, Fayetteville, Arkansas / Luca Capogna, Loredana Lanzani, editors.

p. cm. — (Contemporary mathematics, ISSN 0271-4132 ; 277)

Includes bibliographical references.

ISBN 0-8218-2745-6 (alk. paper)

1. Harmonic analysis. 2. Boundary value problems. I. Capogna, Luca, 1966– II. Lanzani, Loredana, 1965– III. Contemporary mathematics (American Mathematical Society) ; v. 277.

QA403 .H217 2001
515′.2433—dc21 2001033397

♾ The paper used in this book is acid-free and falls within the guidelines
established to ensure permanence and durability.
Visit the AMS home page at URL: http://www.ams.org/

10 9 8 7 6 5 4 3 2 1 06 05 04 03 02 01

Contents

Preface

In this volume we have collected, in final form, some of the papers presented at the 25th University of Arkansas Spring Lecture Series, held in Fayetteville, Arkansas on March 2-4, 2000. The conference was titled *"Recent Progress in the study of harmonic measure from a Geometric and Analytic Point of View"*, and professor C. E. Kenig (University of Chicago) was the principal speaker. Kenig's lectures were complemented by talks from ten invited speakers chosen among leading researchers in the field and twelve junior participants. The present volume includes most of the talks by the former and two selected from the latter group. For space consideration, it was not possible to represent all participants in this publication.

Most of the papers are original research articles. For the few that are not, the authors have strived to give an engaging, well-written, comprehensive survey of their most recent work.

We would like to thank the participants for their contributions, which reflect the quality and the substance of the conference. We are especially grateful to Carlos Kenig, both for his beautiful lectures and for raising many points of discussion during the conference. Kenig's lectures will appear in a separate volume, also published by the AMS.

We are grateful to the University of Arkansas and the National Science Foundation for the financial support which made the conference possible.

A number of people contributed their time and energy to make the conference a success, in particular we wish to thank William Feldman, Collis Geren, Linda Ingels and Itrel Monroe. We also wish to thank Christine Thivierge of the AMS for her invaluable assistance in the editing of these proceedings.

About the Arkansas Spring Lecture Series.

The University of Arkansas Spring Lecture series are yearly conferences organized by the Department of Mathematics of the University of Arkansas.

Each conference is focused on a specific topic chosen from current leading research areas in Mathematics; a principal lecturer delivers a five-lectures course and a number of specialists are invited to give talks on subjects closely related to the topic of the conference.

Short talks by young PhD's and finishing graduate students are solicited to complement the conference.

The Arkansas Spring Lecture Series has grown into an ideal opportunity for specialists and young researchers to meet and exchange ideas about problems at the forefront of modern mathematics.

Over the years, the Arkansas Spring Lectures have been sponsored by the NSF jointly with the University of Arkansas.

Further information is available at:

http://www.uark.edu/depts/mathinfo/Activities/SpringLecture/History.html

Contemporary Mathematics
Volume **277**, 2001

The Mixed Boundary Problem in L^p and Hardy spaces for Laplace's Equation on a Lipschitz Domain

Jeffery D. Sykes and Russell M. Brown

ABSTRACT. We study the boundary regularity of solutions of the mixed problem for Laplace's equation in a Lipschitz graph domain Ω whose boundary is decomposed as $\partial\Omega = N \cup D$, where $N \cap D = \emptyset$. For a subclass of these domains, we show that if the Neumann data g is in $L^p(N)$ and if the Dirichlet data f is in the Sobolev space $L^{p,1}(D)$, for $1 < p < 2$, then the mixed boundary problem has a unique solution u for which $N(\nabla u) \in L^p(\partial\Omega)$, where $N(\nabla u)$ is the non-tangential maximal function of the gradient of u.

1. Introduction

In this paper, we consider the mixed boundary value problem for Laplace's equation in a domain $\Omega \subset \mathbf{R}^n$, $n \geq 3$. We assume that the boundary of Ω is decomposed as $\partial\Omega = N \cup D$, where $N \cap D = \emptyset$. The mixed problem is stated as follows. Given functions f_N and f_D, find a function u which satisfies

$$(1) \qquad \begin{cases} \Delta u = 0 & \text{in } \Omega, \\ u = f_D & \text{on } D, \\ \frac{\partial u}{\partial \nu} = f_N & \text{on } N, \end{cases}$$

where $\frac{\partial u}{\partial \nu} = \nabla u \cdot \nu$ represents the outer normal derivative on $\partial\Omega$. Problems of this kind, including those with elliptic operators more general than the Laplacian and with first order data more general than the normal derivative, have been studied for a long time from many points of view. It is well known that solutions exist under mild conditions on the domain and the data. However, it is also known that the solutions of the more general mixed problems are not generally smooth, regardless of how regular the data may be. In particular, problems arise in a neighborhood of the boundary between N and D. The typical counterexample is given by the harmonic function

$$u(x, y) = \operatorname{Im}(x + iy)^{1/2}, \quad x \in \mathbf{R}, \, y > 0,$$

which is zero on the positive real line and whose normal derivative is zero on the negative real line.

1991 *Mathematics Subject Classification.* Primary 35J25.

Both authors received partial support from the National Science Foundation, Division of Mathematical Sciences.

In results dealing with classical solutions, much work has focused on the conditions necessary to obtain more regular solutions. In **[15]** and **[16]**, Lieberman addresses the Hölder continuity of classical solutions of the mixed problem with smooth data, proving optimal regularity results under certain conditions requiring compatibility between the elliptic operator and the boundary decomposition.

As another example, Azzam and Kreyszig **[1]** consider the mixed problem for elliptic equations in a plane domain with corners. Assuming that the Dirichlet data has Hölder continuous second derivatives and the remaining data has Hölder continuous first derivatives, they prove that the solution has Hölder continuous second derivatives, provided certain conditions are met which relate the interior angle at a corner with the coefficients of the operator at that corner. Their paper also provides a good history of the mixed problem on both smooth domains and domains with corners on the boundary.

Much effort has also been dedicated to the study of the mixed problem in polygonal domains. In his monograph, Grisvard **[12]** addresses both generalized solutions (Chapter 4) and classical solutions (Chapter 6), proving existence and regularity for a larger class of spaces than we consider in our results. This is to be expected, however, as we consider a larger class of domains.

A more recent result is that of Giuseppe Savaré **[19]** who considers the mixed problem in smooth ($C^{1,1}$) domains and who shows that with sufficiently regular data, the solution lies in the Besov space $B^{2,3/2}_\infty(\Omega)$. This addresses one of the limiting cases of Brown's paper **[3]** where it is shown that if we consider the creased domains defined below, then we obtain a non-tangential maximal function estimate which implies that the solution lies in $B^{2,3/2}_2(\Omega)$. The argument connecting the nontangential function estimates and the Besov space results requires the square function studied by Dahlberg **[7]** and the arguments in Fabes's paper **[10]**. Strictly speaking, there is no overlap between the results of Brown and Savaré since no smooth domain satisfies the hypotheses in Brown's results and Savaré requires additional smoothness hypotheses on the data and the domain, compared to the results of Brown.

Since we are considering harmonic functions, our solutions are smooth in the interior. Our interest is in obtaining regularity on the boundary. In particular, we consider a specific class of Lipschitz domains and show that the solution has $\nabla u \in L^p(\partial\Omega)$ when the data f_N is in $L^p(N)$ and f_D is in the Sobolev space $L^{p,1}(D)$, $1 < p < 2$. This result requires, roughly speaking, that D and N meet at an angle which is strictly less than π. See Section 2 for the precise hypotheses.

Our estimates for solutions of the mixed problem will involve the nontangential maximal function. For a function w defined on Ω, the nontangential maximal function of w, denoted $N(w)$, is defined for $y \in \partial\Omega$ by

$$N(w)(y) = \sup_{x\in\Gamma_\alpha(y)} |w(x)|,$$

where for $\alpha > 0$, the nontangential approach region $\Gamma_\alpha(y)$ is defined as

$$\Gamma_\alpha(y) = \{x \in \Omega : |x - y| < (1+\alpha)\delta(x)\}.$$

In this definition and below, we are using $\delta(x) = \text{dist}\,(x, \partial\Omega)$ to denote the distance to the boundary. We note that the dependence of $N(w)$ on α is not significant. In particular, nontangential maximal functions defined using different values of α have comparable L^p-norms (see **[24]**, page 367, for example). For the purposes of this discussion, we require that, in the graph domains defined below, α be chosen so that $y+te_n \in \Gamma_\alpha(y)$ for any $t>0$. Here e_n is the unit vector in the x_n direction.

Boundary values are defined in the nontangential sense: we say that $u=f$ on $\partial\Omega$ if for a.e. $y \in \partial\Omega$,

$$\lim_{\substack{x\to y\\ x\in\Gamma_\alpha(y)}} u(x) = f(y).$$

Except for a set of measure zero, the existence of this limit is independent of the cone opening α. By the outer normal derivative $\frac{\partial u}{\partial \nu}(y)$, we mean $\nu(y)\cdot\nabla u(y)$, where $\nabla u(y)$ is defined nontangentially as above. If Ω is a Lipschitz graph domain and u is harmonic on Ω with $N(u)<\infty$ a.e. on $\partial\Omega$, then u is known to have nontangential limits a.e. on $\partial\Omega$ [**9**].

Our goal is to prove an L^p estimate for solutions of (1):

$$\|N(\nabla u)\|_{L^p(\partial\Omega)} \le C\left(\|f_D\|_{L^{p,1}(D)} + \|f_N\|_{L^p(N)}\right).$$

For $1<p<2$, this first appeared in the Ph.D. thesis of Jeffery Sykes [**23**]. For $p=2$, this result is due to Brown [**3**]. His methods are similar to those used by Jerison and Kenig [**13**] for the Dirichlet and Neumann problems, but Brown makes a new application of the Rellich identity. In particular, he uses a smooth vector-field α so that $|\alpha\cdot\nu| \ge \delta > 0$ a.e. on $\partial\Omega$, but $\alpha\cdot\nu$ changes sign as we cross the boundary between D and N. This can only happen when the normal is discontinuous, which helps explain our hypotheses on $\partial\Omega$, N, and D. This work provides a partial answer to problem 3.2.15 in Kenig's CBMS Lecture Notes [**14**].

In Section 2, we define the class of domains under consideration. The remainder of Section 2 concentrates on proving an L^1 estimate for the L^2 nontangential solution when the Neumann data is an atom for a Hardy space and the Dirichlet data is zero. To accomplish this, we mimic the work of Dahlberg and Kenig [**8**], reflecting the solution in Ω to obtain a solution in a neighborhood of infinity of an elliptic divergence form operator. This allows us to use the asymptotic expansion of Serrin and Weinberger [**20**] to prove that the solution decays at infinity faster than a fundamental solution. This and the same use of the Rellich identity as mentioned above then leads to the atomic estimate. In Section 3, we prove the uniqueness of solutions to the mixed problem with $N(\nabla u)\in L^p(\partial\Omega)$. This is proven by duality, using solutions to the mixed problem with nice data. In Section 4, we give the final steps of the proof of the L^p-result.

We follow the standard convention that C is a constant which varies. Since we are working on graph domains, the constants will depend only on the dimension and the number M which appears below.

2. Atomic Estimates

We begin by defining our problem more precisely. For any point $x \in \mathbf{R}^n$, $n \ge 3$, we denote $x=(x',x_n)=(x_1,x'',x_n)$, where $x_1, x_n \in \mathbf{R}$, $x'' \in \mathbf{R}^{n-2}$, and $x'=(x_1,x'')\in\mathbf{R}^{n-1}$. Let $\Omega=\{x : x_n > \phi(x')\}$ be a Lipschitz graph domain, and let $\psi:\mathbf{R}^{n-2}\to\mathbf{R}$ be a Lipschitz function. Define

$$\begin{array}{rcl} N &=& \{x : x_1 \ge \psi(x'')\}\cap\partial\Omega \\ D &=& \{x : x_1 < \psi(x'')\}\cap\partial\Omega. \end{array} \tag{2}$$

We assume there is a constant M so that ϕ and ψ satisfy

$$\|\nabla\phi\|_{L^\infty(\mathbf{R}^{n-1})} \le M, \qquad \text{and} \qquad \|\nabla\psi\|_{L^\infty(\mathbf{R}^{n-2})} \le M.$$

We also assume that there exist $\delta_D \geq 1/M$ and $\delta_N \geq 1/M$ so that ϕ satisfies

$$\frac{\partial \phi}{\partial x_1} \geq \delta_N \text{ a.e. on } \{x' : x_1 > \psi(x'')\} \tag{3}$$

$$\frac{\partial \phi}{\partial x_1} \leq -\delta_D \text{ a.e. on } \{x' : x_1 < \psi(x'')\}. \tag{4}$$

These conditions are due to Brown **[3]**, whose paper provides an example that illustrates their necessity. Let $\zeta = \{x \in \partial\Omega : x_1 = \psi(x'')\}$ be the boundary between N and D. We will refer to ζ as the "crease" of $\partial\Omega$ and we will call the graph domains we have described *creased domains*.

We will denote by ϕ_N the restriction of ϕ to $\{x' : x_1 \geq \psi(x'')\}$, and ϕ_D will denote the restriction of ϕ to $\{x' : x_1 < \psi(x'')\}$. In the construction below, we will need to have ϕ_D and ϕ_N defined on all of $\mathbf{R}^{n-1}$ so that the conditions (3) and (4) still hold. This is easily accomplished by setting

$$\begin{aligned} \phi_D(x_1, x'') &= \phi_D(\psi(x''), x'') - \delta_D(x_1 - \psi(x'')), \qquad & x_1 \geq \psi(x'') \\ \phi_N(x_1, x'') &= \phi_D(\psi(x''), x'') + \delta_N(x_1 - \psi(x'')), \qquad & x_1 \leq \psi(x''). \end{aligned}$$

In order to lighten the notation, we use the same notation for these extensions to $\mathbf{R}^{n-1}$. Note that with these definitions, we have $\phi = \max(\phi_D, \phi_N)$.

LEMMA 2.1. *For each x_1, the equation $x_1 = \phi_D(t, x'')$ has a unique solution and if we let $t = h(x_1, x'')$ denote the solution, then function h is Lipschitz on $\mathbf{R}^{n-1}$ and the Lipschitz constant depends only on M and δ_D.*

PROOF. Because of our assumption (4), the function $t \to \phi_D(t, x'')$ is strictly monotone and maps $\mathbf{R}$ onto $\mathbf{R}$. Hence, the existence of h follows. If we differentiate the equation

$$x_1 = \phi_D(h(x_1, x''), x'')$$

formally, we obtain

$$\begin{aligned} 1 &= \frac{\partial \phi_D}{\partial x_1}(h(x_1, x''), x'')\frac{\partial h}{\partial x_1}(x_1, x'') \\ 0 &= \frac{\partial \phi_D}{\partial x_1}(h(x_1, x''), x'')\frac{\partial h}{\partial x_j}(x_1, x'') + \frac{\partial \phi_D}{\partial x_j}(h(x_1, x''), x''), \quad j = 2, \ldots, n-2. \end{aligned}$$

Since we have $\partial\phi_D/\partial x_1 \leq -\delta_D$, we can solve for the gradient of h and obtain $\|\nabla h\|_{L^\infty(\mathbf{R}^{n-1})} \leq \max(M, 1)/\delta_D$. If ϕ_D were C^1 and not just Lipschitz, then the implicit function theorem would imply that h is differentiable and the formal calculation would imply that ∇h is bounded.

In the Lipschitz case, a little more work is needed. We regularize the function ϕ_D with a standard mollifier to obtain $\phi_{D,\varepsilon} = \phi_D * \eta_\varepsilon$. If we also assume that $\eta_\varepsilon \geq 0$, then we have that $\phi_{D,\varepsilon}$ also satisfies

$$\frac{\partial \phi_{D,\varepsilon}}{\partial x_1} \leq -\delta_D.$$

We let h_ε be the solution of $\phi_{D,\varepsilon}(h_\varepsilon(x_1, x''), x'') = x_1$. Since $\phi_{D,\varepsilon}$ is smooth, the argument above allows us to conclude that $\|\nabla h_\varepsilon\|_{L^\infty(\mathbf{R}^{n-1})}$ is bounded. We claim that h_ε converges to h uniformly as $\varepsilon \to 0^+$. To establish this claim, we observe that because ϕ_D is Lipschitz, then $\|\phi_{D,\varepsilon} - \phi_D\|_\infty \leq M\varepsilon$. Using this, the monotonicity of $\phi_{D,\varepsilon}$ and the definition of h, we obtain

$$\phi_{D,\varepsilon}(h(x_1,x'')+t,x'') \le x_1 + M\varepsilon - \delta_D t, \qquad t>0$$
$$\phi_{D,\varepsilon}(h(x_1,x'')+t,x'') \ge x_1 - M\varepsilon + \delta_D t, \qquad t<0.$$

Thus, we can conclude that $h_\varepsilon(x_1,x'')$ must satisfy

$$|h(x_1,x'') - h_\varepsilon(x_1,x'')| \le M\varepsilon/\delta_D.$$

Since, we have h_ε converges to h uniformly and $\|\nabla h_\varepsilon\|_{L^\infty(\mathbf{R}^{n-1})}$ is bounded independently of ε, it follows that h is Lipschitz. □

Using the function h, we define a reflection in the graph of ϕ_D by

$$R_1(x_1,x'',x_n) = (2h(x_n,x'')-x_1,x'',x_n).$$

As the figure indicates, this maps acts by mapping a point which is s units to

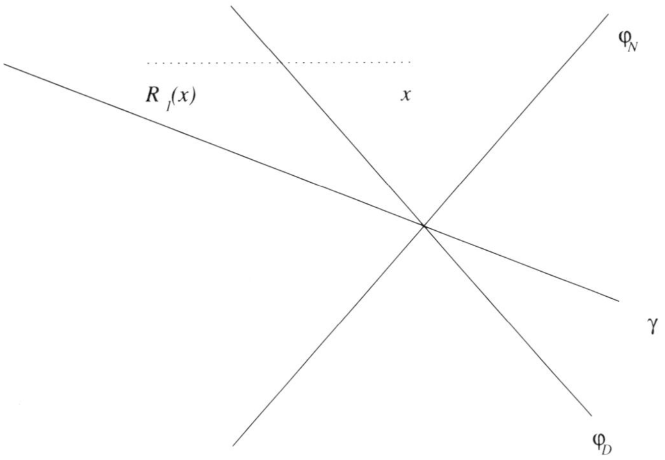

the right of the graph of ϕ_D to a point which is s units to the left of the graph of ϕ_D. It is clear that R_1 is Lipschitz and that $R_1 \circ R_1$ is the identity, hence R_1 is bi-Lipschitz. Next, we define $\Omega_1 = R_1(\Omega) \cup D \cup \Omega$. We claim that the image of the graph $\{(x',x_n) : x_n = \phi_N(x')\}$ is a Lipschitz graph on all of $\mathbf{R}^{n-1}$.

To see this, we would like to apply the implicit function theorem to write the set $\{(x',x_n) : x_n = \phi_N(R_1'(x',x_n))\}$ in the form $\{(x',x_n) : x_n = \gamma(x')\}$ (see figure above). As in Lemma 2.1, we cannot do this directly. Instead, we regularize, apply the implicit function theorem and take a limit. The details are omitted, see [**23**]. Now, we can write $\Omega_1 = \{(x',x_n) : x_n > \phi_1(x')\}$ where the function $\phi_1 = \max(\phi_N,\gamma)$.

Next, we define R_2 to be a reflection in $\partial\Omega_1$ by setting

$$R_2(x',\phi_1(x')+t) = (x',\phi_1(x')-t).$$

Using these reflections, we can change variables to obtain a divergence form operator $L = \operatorname{div} A\nabla$ with bounded measurable coefficients so that $\Delta u = 0$ in Ω if and only if $Lu \circ R_1 = 0$ in $R_1(\Omega)$ and so that $Lu = 0$ in Ω_1 if and only if $Lu \circ R_2 = 0$ in $R_2(\Omega)$. A similar construction is used by Dahlberg and Kenig, see [**8**] or [**23**] for details.

Finally, we introduce notation for function spaces. By $L^p(N)$, we mean the standard space of functions f with $|f|^p$ integrable with respect to surface measure on N. We denote by $L^{p,1}(D)$ the Sobolev space of functions with one derivative in $L^p(D)$. More precisely, $f \in L^{p,1}(D)$ if $f(x', \phi(x')) \in L^{p,1}(\{x' : x_1 < \psi(x'')\})$. We use the norm $\|f\|_{L^{p,1}(\partial\Omega)} = \|\nabla_{tan} f\|_{L^p(\partial\Omega)}$ on $L^{p,1}(D)$. Since we only consider $p \leq 2$ and $n \geq 3$, by Sobolev embedding on the $n-1$ dimensional boundary, we can choose a representative which vanishes at infinity, except when $n = 3$ and $p = 2$. In this case, we will only have uniqueness up to a constant.

In this section, we will prove an L^1 estimate for the nontangential maximal function of the gradient of the solution when the Neumann data is an atom and the Dirichlet data is zero. Recall that a bounded function a is called a $\partial\Omega$-atom if a is supported on a surface ball $B = B_r(y) \cap \partial\Omega = \{x \in \partial\Omega : |x - y| < r\}$ such that $\|a\|_{L^\infty(\partial\Omega)} \leq 1/\sigma(B)$ and $\int a \, d\sigma = 0$.

We call a function a an N-atom if $a = \bar{a}|_N$, where $\bar{a}$ is a $\partial\Omega$-atom. Note that an N-atom a may not have mean value 0, so a is not necessarily a $\partial\Omega$-atom. We then define the N-atomic Hardy space as the ℓ^1-span of the atoms a_j,

$$H^1_{at}(N) = \{f : f = \sum \lambda_j a_j \text{ with } \sum |\lambda_j| < \infty\}$$

with the norm of $f \in H^1_{at}(N)$ defined as

$$\|f\|_{H^1_{at}(N)} = \inf\{\sum |\lambda_j| : f = \sum \lambda_j a_j\}.$$

By this definition, $f \in H^1_{at}(N)$ if and only if $f = \bar{f}|_N$, where $\bar{f} \in H^1_{at}(\partial\Omega)$, and it also follows that

$$\inf\{\|\bar{f}\|_{H^1_{at}(\partial\Omega)} : f = \bar{f}|_N, \bar{f} \in H^1_{at}(\partial\Omega)\}$$

provides an equivalent norm on $H^1_{at}(N)$. See the papers of Chang, Krantz, and Stein [**6**] and Chang, Dafni, and Stein [**5**] for further discussion of defining Hardy spaces on domains in $\mathbf{R}^n$.

We consider the mixed problem with zero Dirichlet data and N-atomic Neumann data, a. Our goal is to show that for such a solution, we have the gradient in $L^1(\partial\Omega)$. Thus the main result of this section is the following proposition.

PROPOSITION 2.2. *Let Ω with N and D be a creased domain. If a is an N-atom and we solve the mixed problem (1) with data $f_D = 0$ and $f_N = a$, then the solution u satisfies*

$$\|N(\nabla u)\|_{L^1(\partial\Omega)} \leq C$$

where the constant C depends only on M.

In order to prove this estimate, we first need to know a solution exists. This is obtained from the earlier work of Brown [**3**].

PROPOSITION 2.3. *If Ω is a creased domain, $f_N \in L^2(N)$ and $\nabla_{tan} f \in L^2(D)$, then there is a unique solution to (1) with $N(\nabla u) \in L^2(\partial\Omega)$.*

PROOF. The existence is obtained by a straightforward extension of the arguments in Brown [**3**]. This paper considers the equation $\Delta - k^2$ for $k \in \mathbf{R} \setminus \{0\}$, the extension to $k = 0$ is straightforward. The uniqueness in graph domains may be proven using the idea of this paper. If we apply the divergence theorem, we obtain that

$$\int_{\partial\Omega} |\nabla u|^2 e_1 \cdot \nu - 2\frac{\partial u}{\partial \nu} e_1 \cdot \nabla u \, d\sigma = 0$$

since $\operatorname{div}(|\nabla u|^2 e_1 - 2\nabla u e_1 \cdot \nabla u) = 0$. This is a version of the Rellich identity, see [**18, 13**]. The condition $N(\nabla u) \in L^2(\partial\Omega)$ is used to justify the integration by parts.

The hypotheses on Ω are designed so that the vector e_1 satisfies $e_1 \cdot \nu > \delta_N/\sqrt{1+M^2}$ a.e. on N and $e_1 \cdot \nu \leq -\delta_D/\sqrt{1+M^2}$ a.e. on D. Rearranging terms and using Cauchy's inequality gives

$$(5) \quad \int_N |\nabla_{tan} u|^2 \, d\sigma + \int_D \left| \frac{\partial u}{\partial \nu} \right|^2 \, d\sigma \leq C \left(\int_D |\nabla_{tan} u|^2 \, d\sigma + \int_N \left| \frac{\partial u}{\partial \nu} \right|^2 \, d\sigma \right).$$

The estimate (5) tells us that that if u has zero data in the mixed problem, then $\nabla u = 0$ on $\partial\Omega$. Now uniqueness in the regularity problem tells us that $u = 0$. See the work of Dahlberg and Kenig [**8**] for uniqueness of the regularity problem in graph domains. □

A key step in the proof of Proposition 2.2 is to consider solutions of a divergence form operator in the complement of a ball. There is an asymptotic expansion for these solutions which allows us to conclude that the solution has some extra decay if a certain integral vanishes. This is an old result of Serrin and Weinberger [**20**] which recall here. If u is a solution of a elliptic divergence form operator div $A\nabla$ in the complement of a ball, $\mathbf{R}^n \setminus B_1(x_0)$, then we have the following asymptotic expansion for u

$$(6) \qquad u(x) = u_\infty + \beta G(x) + w(x)$$

where u_∞ is a constant. The function G is the green's function with pole at x_0 which is known to satisfy $G(x) \leq C|x - x_0|^{2-n}$. The constant β is the outflow which is given by

$$\beta = \int_{\mathbf{R}^n} A(x)\nabla u(x) \cdot \nabla \psi(x) \, dx$$

where ψ is any Lipschitz function which is one in a neighborhood of ∞ and zero in a neighborhood of $\{x : |x - x_0| \leq 1\}$. Finally, v is a function which satisfies

$$|v(x)| \leq C|x - x_0|^{2-n-\delta} \int_{2<|x-x_0|<3} |u(x)| \, dx$$

for constant C and $\delta > 0$ which depend only on the dimension and the bounds on the eigenvalues of the matrix of coefficients A.

PROOF. Now we turn to the proof of Proposition 2.2. We consider an N-atom a. Since the estimate we desire to prove and the class of creased domains are invariant under dilation, we may assume that the ball appearing the definition of an N-atom has radius $r = 1$. According to Proposition 2.3, there is a solution u to (1) with $f_D = 0$ and $f_N = a$ and with

$$\|N(\nabla u)\|_{L^2(\partial\Omega)} < \infty.$$

Our goal is to show that ∇u decays sufficiently fast at infinity so that $N(\nabla u) \in L^1(\partial\Omega)$. We extend u by odd reflection about D using the map R_1 defined above to obtain u in Ω_1, satisfying

$$\begin{aligned} u_1 \circ R_1 &= -u_1 \\ Lu_1 &= 0 \end{aligned}$$

for the divergence form elliptic operator L we defined above. Since $N(\nabla u) \in L^2(\partial\Omega)$, it follows that $\nabla u_1 \in L^2(B)$ for any bounded set $B \subset \Omega_1$ and thus that

u_1 is a weak solution of the Neumann problem

$$\begin{cases} Lu = 0 & \text{in } \Omega_1 \\ A\nabla u \cdot \nu = a + a^* & \text{on } \partial\Omega_1 \end{cases}$$

where $a^*(y) = -(DR_1 \circ R_1^{-1})^t \nu(y) \cdot \nu(R_1^{-1}(y)) a(R_1(y))$. Since we do not yet know that ∇u is in $L^2(\Omega)$, we require that the test functions in the weak formulation of this problem be zero outside some large ball. However, we do not require test functions to vanish on $\partial\Omega$. Changing variables, we see that

$$\int_N a(y)\, d\sigma(y) = -\int_{R_1(N)} a^*(y)\, d\sigma(y).$$

Thus, if dist (supp $a, D) < 1$, then $\lambda(a+a^*)$ is an atom for $\partial\Omega_1$, where the constant λ depends only on M. If dist (supp $a, D) > 1$, then we have that a is a $\partial\Omega_1$ atom and λa^* is a multiple of a $\partial\Omega_1$ atom for λ depending only on permissible quantities.

We may reflect in $\partial\Omega_1$ using R_2 to obtain an even extension u_2 which solves $Lu_2 = 0$ in $\mathbf{R}^n \setminus \text{supp}\,(a+a^*)$. Applying the theorem of Serrin and Weinberger (6),

$$u_2(x) = \beta g(x) + w(x) \tag{7}$$

where $|w(x)| = O(|x|^{2-n-\delta})$ for some $\delta > 0$. We have that $u_\infty = 0$ since u vanishes on D. Now the estimate $|u_2(x)| \le C|x|^{2-n}$ for x large and Caccioppoli's inequality implies $\nabla u \in L^2(\mathbf{R}^n)$, $n \ge 3$. Thus we can conclude that the solution u_2 satisfies

$$\|\nabla u_2\|_{L^2(\mathbf{R}^n)} < \infty. \tag{8}$$

Our next step is to claim that there exist $C > 0$ and $\delta > 0$ so that

$$|u(x)| \le C|x|^{2-n-\delta}, \qquad x \in \Omega. \tag{9}$$

In this estimate, the constant C depends only on the constant M. In order to obtain the correct dependence of the constant, we must be more careful than we were in the previous paragraph. This requires consideration of the cases dist (supp $a, D) \le 1$ and dist (supp $a, D) > 1$ separately.

We first consider dist (supp $a, D) \le 1$. We recall that in the expansion (7), $C_0 = 0$ since $u = 0$ on D and

$$\beta = \int_{\mathbf{R}^n} A(x)\nabla\eta(x) \cdot \nabla u_2(x)\, dx$$

where η is any function which is zero on the support of the Neumann data a and is identically one for all x sufficiently large.

As in Dahlberg and Kenig [**8**, p. 444], we have

$$\int_{\mathbf{R}^n} A(x)\nabla\eta(x)\nabla u(x)\, dx = 2\int_{\partial\Omega_1} (a(x) + a^*(x))\, d\sigma(x) = 0$$

and hence $\beta = 0$. Now, we can estimate u in the set of points which are at most distance 2 from the support of a using the $L^2(\partial\Omega)$ estimate for $N(\nabla u)$. Since u vanishes on D, a Poincaré inequality gives us an estimate for the integral of $|u|^2$ in Ω near support a. This gives the estimate (9) with correct dependence of the constant.

If dist (supp $a, D) > 1$, then the above procedure gives the correct order at infinity, but the constant C depends on the distance between the support of a and

a^*. In this case, we need to split the solution $u_1 = u_{11} + u_{12}$ where u_{11} is a weak solution of the Neumann problem

$$\begin{cases} Lu_{11} = 0, & \text{in } \Omega_1 \\ \frac{\partial u_{11}}{\partial \nu} = a, & \text{on } \partial\Omega_1. \end{cases}$$

The function u_{12} solves the same problem with data a^* instead of a. We claim these solutions each satisfy the estimates

$$\begin{aligned} (10)\ |u_{11}(x)| &\leq C \operatorname{dist}(x, \operatorname{supp} a)^{2-n-\delta}, && x \in \Omega_1,\ \operatorname{dist}(x, \operatorname{supp} a) > 1 \\ (11)\ |u_{12}(x)| &\leq C \operatorname{dist}(x, \operatorname{supp} a^*)^{2-n-\delta}, && x \in \Omega_1,\ \operatorname{dist}(x, \operatorname{supp} a^*) > 1 \end{aligned}$$

Together, (10) and (11) imply the estimate (9) in the second case.

The proof of (10) and (11) depends on reflecting in Ω_1 and then using the Serrin-Weinberger theorem (6). Again, we have $u_\infty = 0$ since the energy solutions are in $L^{2n/(2n-2)}(\mathbf{R}^n)$ and $\beta = 0$ since a and a^* have mean value zero. The dependence of the constant (and the existence of u_{11} and u_{12}) depends on the following boundary Poincaré inequality. In this Lemma, we use $⨍_E$ to denote the average over E.

LEMMA 2.4. *If Ω is a Lipschitz graph domain, u is a function with ∇u in $L^2(\Omega)$, $x \in \partial\Omega$ and $r > 0$, then with*

$$\bar{u} = ⨍_{B_r(x)\cap\partial\Omega} u(x)\, d\sigma$$

we have that

$$\int_{\partial\Omega\cap B_r(x)} |u(x) - \bar{u}|^2\, d\sigma(x) \leq Cr^2 \int_{\Omega\cap B_r(x)} |\nabla u(x)|^2\, dx.$$

As a consequence of this Lemma, we can show that the map

$$u \longrightarrow \int_{\partial\Omega_1} au\, d\sigma$$

is a continuous linear functional on the Sobolev space $\dot{L}^{2,1}(\Omega_1)$ which we norm with $(\int_{\Omega_1} |\nabla u|^2\, dx)^{1/2}$. Thus, we can use Lax-Milgram to show that there is a weak solution of the Neumann problem for L with data a and that this solution satisfies $\|\nabla u\|_{L^2(\Omega_1)} \leq C$ (the constant depends on r in general, but here we have no r dependence since we have already set $r = 1$). Then, Sobolev embedding implies that the solution lies in $L^{2n/(2n-2)}(\Omega_1)$ and this can be used to show that the constants in (10) and (11) depend only on the Lipschitz constant M. We also note that the estimate (8) and uniqueness in the class of solutions with ∇u in $L^2(\Omega_1)$ is needed to show $u_1 = u_{11} + u_{12}$.

Now from the estimate (9) we can use Caccioppoli's inequality to find that on the ring

$$R_k = \{x : 2^k \leq \operatorname{dist}(x, \operatorname{supp} a) < 2^{k+1}\},$$

we have

$$(12) \qquad \int_{R_k\cap\Omega} |\nabla u|^2\, dx \leq C2^{k(2-n-2\delta)}, \qquad k = 0, 1, \ldots$$

for some constant $C = C(M)$ and for δ as in the Serrin-Weinberger theorem (see (9)).

Our next step is to obtain estimates for ∇u on the boundary. We let $\eta \geq 0$ be a cutoff function with $\eta = 1$ on R_k and $\operatorname{supp} \eta \subset \cup_{|j|\leq 1} R_{k+j}$ and $|\nabla\eta| \leq C2^{-k}$.

We again use the Rellich identity as in the proof of Proposition 2.3, but with e_1 replaced by ηe_1 to obtain that

$$\begin{aligned}\int_{R_k\cap\partial\Omega} |\nabla u(x)|^2\,d\sigma(x) &\leq C\int_{\partial\Omega} \eta(x)|\nabla u(x)|^2\,d\sigma(x)\\ &\leq C\int_{\Omega} |\nabla\eta(x)||\nabla u(x)|^2\,dx.\end{aligned}$$

Now, Caccioppoli's inequality (12) and our observation that u has mixed data 0 away from supp a gives that

$$\int_{R_k\cap\partial\Omega} |\nabla u(x)|^2\,dx \leq C2^{k(1-n-2\delta)}, \qquad k\geq 1.$$

Finally, the localization argument as in Dahlberg and Kenig [**8**, p. 445] gives

$$\int_{R_k} N(\nabla u)(x)^2\,d\sigma(x) \leq C2^{k(1-n-2\delta)}.$$

This estimate implies that we have

$$\int_{\partial\Omega} N(\nabla u)(x)\,d\sigma(x) \leq C.$$

This concludes the proof of the existence of solutions as described in Proposition 2.2. □

Finally, we give the main result of this section. The existence of solutions with atomic data. Before we can do this, we must define the space $H^{1,1}_{at}(D)$, the space of functions with one derivative in the Hardy space. The simplest way to define this is to observe that, like $H^1(N)$, $H^{1,1}_{at}(D)$ is the restriction to D of the corresponding space on $\partial\Omega$. Thus, we say b is a 1-atom for $\partial\Omega$ if b is supported in a ball $B_r(x)\cap\partial\Omega$ and $\|\nabla_{tan}b\|_{L^\infty(B_r(x)\cap\partial\Omega)} < r^{2-n}$. As before, b is 1-atom for D if $b=\bar{b}|_D$ for $\bar{b}$ a 1-atom for $\partial\Omega$. Finally, the space $H^{1,1}_{at}(D)$ is defined as the ℓ^1-span of 1-atoms for D.

THEOREM 2.5. *Let Ω, N and D give a creased domain. If f_N is in $H^1(N)$ and f_D is in $H^{1,1}_{at}(D)$, then there exists a solution of the mixed problem (1) which satisfies*

$$\|N(\nabla u)\|_{L^1(\partial\Omega)} \leq C(\|f_N\|_{H^1_{at}(N)} + \|f_D\|_{H^{1,1}_{at}(D)}).$$

Furthermore, if the f_N is also in $L^2(N)$ and f_D is also in $L^{2,1}(D)$, then the solution satisfies $N(\nabla u)\in L^1(\partial\Omega)\cap L^2(\partial\Omega)$.

PROOF. The result for Hardy spaces follows from Proposition 2.2. To see this, we first use the result of Dahlberg and Kenig on the regularity of the solution of the Dirichlet problem when the Dirichlet data is in space $H^{1,1}_{at}(\partial\Omega)$. This result allows us to reduce the case when f_D is zero. For this special case, we can write the data f_N as a sum of atoms and then add the solutions from Proposition 2.2.

The additional statements about L^2, follow from the observation of Pipher and Verchota [**17**] that if we take a $L^2\cap H^1$ function, then the atomic decomposition can be chosen so that the sum converges in L^2 also. This technical observation is needed in the interpolation argument in section 4. □

3. Uniqueness

Now we turn to uniqueness. This seems to be rather involved. The proof for uniqueness in the L^p problem is the same as for the Hardy space problem, thus we do them simultaneously.

PROPOSITION 3.1. *If Ω is a creased domain, $N(\nabla u) \in L^p(\partial\Omega)$, $1 \leq p < 2$, $n \geq 3$ and u is a solution of the mixed problem with $f_D = 0$ and $f_N = 0$, then $u = 0$.*

PROOF. The argument proceeds by duality. Thus, we consider a solution w of the mixed problem with Dirichlet data g_D, a Lipschitz function with compact support in $\mathbf{R}^n$ and zero Neumann data. Thus w satisfies:

$$\begin{cases} \Delta w = 0, & \text{in } \Omega \\ w = g_D, & \text{on } D \\ \frac{\partial w}{\partial \nu} = 0, & \text{on } N \end{cases}$$

The function g_D will be a multiple of a 1-atom on D for the space $H^{1,1}_{at}(D)$. A straightforward adaptation of the argument for existence of solutions with atomic Neumann data allows us to show that the solution w satisfies

$$N(\nabla w) \in L^1(\partial\Omega) \cap L^2(\partial\Omega). \tag{13}$$

Next, we observe that the solution w lies in the Hölder space $C^\alpha(\bar{\Omega})$ for some $\alpha = \alpha(M) > 0$ and thus

$$\|u\|_{C^\alpha(\bar{\Omega})} = \sup_{x \neq y, x, y \in \Omega} \frac{|w(x) - w(y)|}{|x - y|^\alpha} \leq C(g_D). \tag{14}$$

The equality serves to define the Hölder semi-norm. The inequality in (14) is probably well-known. It can be proven by reflecting w in the Neumann face N and using the Hölder continuity at the boundary of weak solutions of the Dirichlet problem (see [**11**]).

Since we have $N(\nabla w)$ is in $L^1(\partial\Omega) \cap L^2(\partial\Omega)$, we can use a version of the Sobolev embedding theorem to conclude that

$$N(w) \in L^{(n-1)/(n-2)}(\partial\Omega) \tag{15}$$

(see Brown [**4**] for a proof).

Next, we recall the (generalized) Riesz transforms on a Lipschitz graph domain. If $N(w) \in L^p(\partial\Omega)$, $p < \infty$, then with $r = \delta(x)/2$, we have

$$|\nabla w(x)| \leq \frac{C}{r} \fint_{B_r(x)} |w(x)|\, dx \leq C\delta(x)^{-1-(n-1)/p} \|N(w)\|_{L^p(\partial\Omega)}.$$

Thus, we define for $j = 1, \ldots, n$

$$w_j(x) = -\int_{x_n}^{\infty} \frac{\partial w}{\partial x_j}(x', t)\, dt. \tag{16}$$

The collection $w_1, \ldots, w_n$ satisfy

$$\begin{aligned} w_n &= w \\ \frac{\partial w_j}{\partial x_k} &= \frac{\partial w_k}{\partial x_j} \qquad j, k = 1, \ldots, n \\ \sum_{j=1}^{n} \frac{\partial w_j}{\partial x_j} &= 0. \end{aligned} \tag{17}$$

The last two equations in (17) are the system for conjugate harmonic functions as studied by Stein and Weiss [**22**]. Using square function arguments as in Stein's book, [**21**], one can show that

$$\|N(w_i)\|_{L^p(\partial\Omega)} \le C\|N(w)\|_{L^p(\partial\Omega)}, \qquad 0<p<\infty. \tag{18}$$

The argument for the half-space as presented by Stein extends immediately to a Lipschitz graph domain, once we have the equivalence of the non-tangential maximal function and the square function as proven by Dahlberg [**7**].

Next, we rely on additional ideas from Littlewood-Paley theory to show that each w_i is also in $C^\alpha(\bar\Omega)$. This is proven using two observations about harmonic functions in Lipschitz graph domains. First, if w is C^α, $0<\alpha\le 1$, and harmonic, then for each multi-index β with $|\beta|>0$,

$$\sup_{x\in\Omega} \delta(x)^{|\beta|-\alpha} \left|\frac{\partial^\beta w}{\partial x^\beta}(x)\right| \le C\|w\|_{C^\alpha(\bar\Omega)}. \tag{19}$$

This is proven by estimating the derivative of $w(\cdot)$ by the average of $w(\cdot)-w(x)$ on a ball of radius half the distance to the boundary. Conversely, if w is differentiable in a Lipschitz graph domain Ω, then

$$\|w\|_{C^\alpha(\bar\Omega)} \le \sup_{x\in\Omega} \delta(x)^{1-\alpha}|\nabla w(x)|. \tag{20}$$

From (14), we have $w=w_n$ is Hölder continuous in Ω and thus the second derivatives of w satisfy the condition (19). If we differentiate the definition of w_i, (16), and use (19) we obtain that the gradient of each w_i satisfies

$$\sup \delta(x)^{1-\alpha}|\nabla w_i(x)| \le C\|w_n\|_{C^\alpha(\bar\Omega)}.$$

Hence, from (20) we have that

$$\|w_i\|_{C^\alpha(\bar\Omega)} \le C(g_D). \tag{21}$$

Next, we observe that if $\|N(w)\|_{L^p(\Omega)}<\infty$ and $\|w\|_{C^\alpha(\bar\Omega)}<\infty$, then w must be bounded. To see this, use the nontangential maximal function to obtain that w is bounded when $\delta(x)>1$ and then the Hölder continuity near the boundary. Applying this observation to each w_i, we conclude that each of the w_i is bounded in Ω. Combining the estimate (21) with the estimates (15) and (18) for each w_i, we obtain

$$\|N(w_i)\|_{L^{(n-1)/(n-2)}(\partial\Omega)} + \|w_i\|_{L^\infty(\bar\Omega)} \le C(g_D). \tag{22}$$

Now, we have u as in Proposition 3.1, a solution of the mixed problem with $N(\nabla u)\in L^p(\partial\Omega)$ and zero data in the mixed problem (1). For $s>0$ we let u_s denote a translation of u defined by $u_s(x)=u(x+se_n)$ and define $w_t(x)=w(x+te_n)$. We consider u_s, w_t and a cutoff function η_R which is one on a ball of radius R and 0 outside a ball of radius $2R$. Applying Green's second identity gives

$$\int_{\partial\Omega} \eta_R\Big(u_t\frac{\partial w_s}{\partial\nu} - w_s\frac{\partial u_t}{\partial\nu}\Big)\,d\sigma = \int_\Omega \nabla\eta_R(x)\cdot(u_t(x)\nabla w_s(x) - w_s(x)\nabla u_t(x))\,dx.$$

We consider, in turn, the various limits as t and s approach zero and R approaches infinity. First, consider R going to infinity. Since $N(\nabla u)$ is in $L^p(\partial\Omega)$, the Sobolev embedding Lemma in [**4**]

$$N(u_t)\in L^{p(n-1)/(n-1-p)}(\partial\Omega), \qquad t\ge 0. \tag{23}$$

Also, we have $N(\nabla w_s) \in L^1(\partial\Omega)$ and for $s > 0$, $|\nabla w_s| \leq C_s$, when $s > 0$. Thus,

$$\int_\Omega |\nabla \eta_R \cdot (u_t \nabla w_s)|\, dx \leq C \int_{\{x:|x|\geq cR\}\cap\partial\Omega} N(u_t)N(\nabla w_s)\, d\sigma$$

which vanishes in the limit as $R \to \infty$ since u satisfies (23) and we have $N(\nabla w_s)$ is in $L^1(\partial\Omega) \cap L^\infty(\partial\Omega)$ (though with a constant which depends on s). Next, we consider

$$\int_\Omega |w_s \nabla \eta_R \cdot \nabla u_t|\, dx \leq C \int_{\{x:|x|>cR\}} N(w_s)N(\nabla u_t)\, d\sigma.$$

Since we have $N(\nabla u_t) \in L^p(\partial\Omega)$ for some p in $[1, 2)$, and $w_s \in L^{(n-1)/(n-2)}(\partial\Omega) \cap L^\infty$, we have that this term also vanishes in the limit as $R \to \infty$. Finally, we may use dominated convergence to find the limit

$$\lim_{R\to\infty} \int_{\partial\Omega} \eta_R(u_t \frac{\partial w_s}{\partial\nu} - w_s \frac{\partial u_t}{\partial\nu})\, d\sigma = 0.$$

This is because (as we have shown above) the expression multiplying η_R is in $L^1(\partial\Omega)$. Thus, we have shown for all $s > 0$ and all $t > 0$ that

$$\int_{\partial\Omega} u_t \frac{\partial w_s}{\partial\nu} - w_s \frac{\partial u_t}{\partial\nu}\, d\sigma = 0. \tag{24}$$

Our next step is to let s and then t go to zero. We begin by showing that

$$\lim_{s\to 0^+} \int_{\partial\Omega} u_t \frac{\partial w_s}{\partial\nu}\, d\sigma = \int_{\partial\Omega} u_t \frac{\partial w}{\partial\nu}\, d\sigma. \tag{25}$$

We take advantage of the fact that for each $t > 0$, u_t is smooth and hence bounded in Ω. The function $N(\nabla w)$ is in $L^1(\partial\Omega)$ and hence (25) follows easily from the dominated convergence theorem. Next, we consider

$$\lim_{t\to 0^+} \int_{\partial\Omega} w_s \frac{\partial u_t}{\partial\nu}\, d\sigma = \int_{\partial\Omega} w_s \frac{\partial u}{\partial\nu}\, d\sigma. \tag{26}$$

This follows since we have $\nabla u \in L^p(\partial\Omega)$ for some p, $1 \leq p < 2$ and w satisfies (22). The same estimates allow us to use the dominated convergence theorem to show

$$\lim_{s\to 0^+} \int_{\partial\Omega} w_s \frac{\partial u}{\partial\nu}\, d\sigma = \int_D g_D \frac{\partial u}{\partial\nu}\, d\sigma. \tag{27}$$

The integral on the right-hand side of this equation is only over D since $\frac{\partial u}{\partial\nu} = 0$ on N.

After these routine limits, we now come to the interesting part of the argument. We claim that

$$\lim_{s\to 0^+} \int_{\partial\Omega} u \frac{\partial w_s}{\partial\nu}\, d\sigma = 0. \tag{28}$$

This presents a challenge since (if say, $p = 1$) we have estimates for u in $L^{(n-1)/(n-2)}(\partial\Omega)$, but we only have $N(\nabla w)$ in $L^p(\partial\Omega)$ for $p \leq 2$.

Our argument relies on a technique from G. Verchota's paper [**25**] which uses the generalized Riesz transforms to express the integrand in (28) in terms of derivatives of u and the boundary values of w_i. These quantities are easily estimated as in our first three limits.

To proceed, we write out the integral over $\partial\Omega$ in the coordinates $(x', \phi(x'))$ and the use the system that $w_1, \ldots, w_n$ satisfies to obtain

$$\begin{aligned}
(29)\quad \int_{\partial\Omega} u\frac{\partial w_s}{\partial\nu}\,d\sigma &= \int_{\mathbf{R}^{n-1}} u(x',\phi(x'))(\sum_{i=1}^{n-1}\phi_{x_i}(x')\frac{\partial w_s}{\partial x_i}(x',\phi(x')) \\
&\quad -\frac{\partial w_s}{\partial x_n}(x',\phi(x')))\,dx' \\
(30)\quad &= \int_{\mathbf{R}^{n-1}} u(x',\phi(x'))(\sum_{i=1}^{n-1}\phi_{x_i}(x')\frac{\partial w_{i,s}}{\partial x_n}(x',\phi(x')) \\
(31)\quad &\quad +\frac{\partial w_{i,s}}{\partial x_i}(x',\phi(x')))\,dx' \\
&= -\int_{\mathbf{R}^{n-1}}\sum_{i=1}^{n-1}\frac{\partial}{\partial x_i}u(x',\phi(x'))w_{i,s}(x',\phi(x'))\,dx'.
\end{aligned}$$

The boundary terms at infinity in the integration by parts vanish since we have $w_i \in L^{(n-1)/(n-2)}(\partial\Omega)\cap L^\infty(\partial\Omega)$ and $u \in L^{p(n-1)/(n-1-p)}(\partial\Omega)$. We omit the details. Now, as above, we may use $w_i \in L^q(\partial\Omega)$, for all q with $(n-1)/(n-2) \leq q \leq \infty$ and that $\nabla u \in L^p(\partial\Omega)$, for some p, $1 \leq p < 2$ to conclude that

$$\lim_{s\to 0^+}\int_{\partial\Omega} w_{i,s}(x',\phi(x'))\frac{\partial}{\partial x_i}u(x',\phi(x'))\,dx' = \int_{\mathbf{R}} w_i(x',\phi(x'))\frac{\partial}{\partial x_i}u(x',\phi(x'))\,dx'.$$

Now we have the task of reversing the calculations in (31) to conclude that this last integral is zero.

To do this, we first observe that, away from the crease, we have some additional regularity of the solution u. This follows from the solution of the Neumann and regularity problem as in Dahlberg and Kenig, [**8**]. In particular, using the Green's function for each of these problems as in [**2**, p. 33] (which is based on [**8**]), we can show that if B_r is a ball centered at a point in $\partial\Omega$ and $B_{2r}\cap\partial\Omega$ is contained in N or in D, but does not meet both sets, then we have

$$(32)\qquad \sup_{y\in B_r(x)\cap\Omega}|u(y)| \leq C\int_{B_{2r}(x)\cap\partial\Omega} N(u)\,d\sigma.$$

Let η_ε be a cutoff function which is one if dist $(x, D) > 2\varepsilon$, is zero when dist $(x, D) < \varepsilon$ and satisfies $|\nabla\eta_\varepsilon| \leq C/\varepsilon$. Also, let $\tilde\eta_R$ be a cutoff function which is one if $|x| < R$, zero if $|x| > 2R$ and satisfies $|\nabla\tilde\eta_R| \leq C/R$. Then since the integrand is in $L^1(\mathbf{R}^{n-1})$,

$$\begin{aligned}
&\int_{\mathbf{R}^{n-1}} w_i(x',\phi(x'))\frac{\partial}{\partial x_i}u(x',\phi(x'))\,dx' \\
&= \lim_{R\to\infty,\varepsilon\to 0^+}\int_{\mathbf{R}^{n-1}}\eta_\varepsilon(x',\phi(x'))\tilde\eta_R(x',\phi(x'))w_i(x',\phi(x'))\frac{\partial}{\partial x_i}u(x',\phi(x'))\,dx'
\end{aligned}$$

since $\nabla u \in L^p$ for some p, $1 \leq p < 2$ and w_i satisfies (22). Note that we can assume the cutoff functions vanish on D since the tangential derivatives of u are already zero there. Now we consider the integral inside the limit with $\varepsilon > 0$ and integrate by parts to obtain

$$\begin{aligned}
&\int_{\mathbf{R}^{n-1}} \tilde{\eta}_R(x',\phi(x'))\eta_\varepsilon(x',\phi(x'))\, w_i(x',\phi(x'))\frac{\partial}{\partial x_i}u(x',\phi(x'))\,dx' \\
&= -\int_{\mathbf{R}^{n-1}} (\tilde{\eta}_R(x',\phi(x'))\eta_\varepsilon(x',\phi(x'))u(x',\phi(x'))\frac{\partial}{\partial x_i}w_i(x',\phi(x')) \\
&\qquad +u(x',\phi(x'))w_i(x',\phi(x'))\frac{\partial}{\partial x_i}(\tilde{\eta}_R(x',\phi(x'))\eta_\varepsilon(x',\phi(x')))\,dx'
\end{aligned}$$

where the boundary term at infinity vanishes because of our estimates on w_i and u. Since the estimate (32) implies u is bounded away from the crease and $N(\nabla w_i)$ is in $L^1(\partial\Omega)$, we may undo the calculation in (31) above to show that

$$\begin{aligned}
&\sum_{i=1}^{n-1}\int_{\mathbf{R}^{n-1}} \frac{\partial}{\partial x_i}w_i(x',\phi(x'))u(x',\phi(x'))\tilde{\eta}_R(x',\phi(x'))\eta_\varepsilon(x',\phi(x'))\,d\sigma \\
&= \int_{\partial\Omega} w\frac{\partial u}{\partial\nu}\tilde{\eta}_R\eta_\varepsilon\,d\sigma = 0.
\end{aligned}$$

Of course, this is because $\partial w/\partial\nu = 0$ on N and $u = 0$ on D. Thus, to finish, we need to show that

$$\begin{aligned}
\lim_{\varepsilon\to 0^+}\int_N |w_i u\frac{\partial}{\partial x_i}\eta_\varepsilon|\,d\sigma &= 0 \\
\lim_{R\to\infty}\int_N |w_i u\frac{\partial}{\partial x_i}\tilde{\eta}_R|\,d\sigma &= 0
\end{aligned}$$

We consider the limit involving ε first. Here, we use that $u = 0$ on D to prove a Poincarè inequality: For some $C > 0$, we have

$$\int_{\{x:\mathrm{dist}\,(x,\zeta)\le 2\varepsilon\}} |u(x)|^p\,d\sigma(x) \le C\varepsilon^p\int_{\{x\in\partial\Omega:\mathrm{dist}\,(x,\zeta)\le C\varepsilon\}} |\nabla u(x)|^p\,d\sigma.$$

Finally, if we let ζ_ε denote $\{x : \mathrm{dist}\,(x,\zeta) < C\varepsilon\}$ then we have

$$\int_N |uw_i\frac{\partial}{\partial x_i}\eta_\varepsilon|\,d\sigma \le \lim_{\varepsilon\to 0^+} C\left(\int_{\zeta_\varepsilon}|\nabla u(x)|^p\,d\sigma\right)^{1/p}\left(\int_{\zeta_\varepsilon}|w_i(x)|^{p'}\right)^{1/p'} = 0.$$

This uses that ∇u is in $L^p(\partial\Omega)$ and w_i is in $L^{p'}(\partial\Omega)$. The limit that $R\to\infty$ vanishes because we have w_i in $L^{(n-1)/(n-2)}(\partial\Omega)$ and $u \in L^{p(n-1)/(n-1-p)}(\partial\Omega)$. This completes the proof of the claim (28).

Combining (24), (25), (26), (27) and (28), we have that $\partial u/\partial\nu$ is zero on $\partial\Omega$. Thus the uniqueness for the Neumann problem proved in [**8**] implies that u is zero. This completes the proof of uniqueness for p in the range $1 \le p < 2$. □

4. Interpolation

Our last step, is to show existence of solutions for p in the range, $1 < p < 2$. This is done by interpolation. Only one or two details are needed.

PROPOSITION 4.1. *Let Ω, N and D be a creased domain. If f_D is in $L^{p,1}(D)$ and $f_N \in L^p(N)$, then there exists a solution u to (1) which satisfies*

$$\|N(\nabla u)\|_{L^p(\partial\Omega)} \le C(\|f_N\|_{L^p(\partial\Omega)} + \|f_D\|_{L^{p,1}(\partial\Omega)}).$$

PROOF. Thanks to the solvability of the Dirichlet problem with data f which has $\nabla_{tan}f \in L^p(\partial\Omega)$ [**8**, Theorem 3.8], we may assume that $f_D = 0$. If f is in $L^p(N)$, $1 < p < 2$, then we may extend f to all of $\partial\Omega$ by setting $f = 0$ on D. We consider the function

$$\tilde{f}(x') = f(x', \phi(x'))\sqrt{1 + |\nabla\phi(x')|^2}.$$

According to the decomposition lemma in the monograph of Torchinsky, [**24**, p. 364], we have for $\lambda > 0$ that we can write

$$\tilde{f} = f_\lambda + f^\lambda$$

where

$$\|f^\lambda\|_{H^1(\mathbf{R}^{n-1})} \leq C\lambda^{1-p}\|f\|_p^p \quad \text{and} \quad \|f_\lambda\|_{L^2(\mathbf{R}^{n-1})} \leq C\lambda^{2-p}.$$

Thus we may use uniqueness in the L^2-mixed problem and the observation about L^2-solutions in Theorem 2.5 to show that

$$u = u^\lambda + u_\lambda$$

where u is the solution of (1) with data $f_N = f|_N$ and $f_D = 0$ and u_λ has Neumann data $f_N(x', \phi(x')) = (1 + |\nabla\phi(x')|^2)^{-1/2} f_\lambda(x')$ and $f_D = 0$. The function u^λ solves a mixed problem with data involving f^λ.

Next, from our estimates for H^1 and L^2, we obtain the weak-type estimate,

$$\sigma(\{x : N(\nabla u) > \lambda\}) \leq \sigma(\{N(\nabla u^\lambda) > \lambda/2\}) + \sigma(\{N(\nabla u_\lambda) > \lambda/2\}) \leq \frac{C}{\lambda^p}\|f\|_p^p.$$

This weak type estimate and the Marcinkiewicz interpolation theorem (see [**21**, p. 272]) give the strong-type inequality. Thus we have that if $f \in L^2 \cap L^p$, there is a solution to (1) which satisfies

$$\|N(\nabla u)\|_{L^p(\partial\Omega)} \leq C\|f\|_{L^p(\partial\Omega)}.$$

To remove the restriction that f is in $L^2 \cap L^p$, involves a routine limiting argument. □

Finally, we summarize what we have done. The result of this theorem is the main result of this paper.

THEOREM 4.2. *Suppose* $1 < p < 2$ *and* $n \geq 3$. *Let* Ω, N *and* D *be a creased domain.*

If f_N *is in* $L^p(N)$ *and* f_D *is in* $L^{p,1}(D)$, *then there exists a unique solution to the mixed problem (1) which satisfies*

$$\|N(\nabla u)\|_{L^p(\partial\Omega)} \leq C(\|f_N\|_{L^p(N)} + \|f_D\|_{L^{p,1}(D)}).$$

If f_N *is in* $H^1_{at}(N)$ *and* f_D *is in* $H^{1,1}_{at}(D)$, *then there exists a unique solution to the mixed problem (1) which satisfies*

$$\|N(\nabla u)\|_{L^1(\partial\Omega)} \leq C(\|f_N\|_{H^1_{at}(N)} + \|f_D\|_{H^{1,1}_{at}(D)}).$$

In each case, the uniqueness assertion is in the class of solutions which have $N(\nabla u)$ *in* $L^p(\partial\Omega)$. *When* $p = 2$ *and* $n = 3$, *we only have uniqueness modulo constants.*

References

[1] A. Azzam and E. Kreyszig. On solutions of elliptic equations satisfying mixed boundary conditions. *SIAM J. Math. Anal.*, 13(2):254–262, 1982.

[2] R. M. Brown. The initial-Neumann problem for the heat equation in Lipschitz cylinders. *Trans. Amer. Math. Soc.*, 320:1–52, 1990.

[3] R. M. Brown. The mixed problem for Laplace's equation in a class of Lipschitz domains. *Comm. Partial Diff. Eqns.*, 19:1217–1233, 1994.

[4] R. M. Brown. The Neumann problem on Lipschitz domains in Hardy spaces of order less than one. *Pac. J. Math.*, 171(2):389–407, 1995.

[5] D. C. Chang, G. Dafni, and E. M. Stein. Hardy spaces, BMO, and boundary value problems for the Laplacian on a smooth domain in $\mathbf{R}^N$. *Trans. Amer. Math. Soc.*, 351(4):1605–1661, 1999.

[6] D.-C. Chang, S. G. Krantz, and E. M. Stein. H^p theory on a smooth domain in $\mathbf{R}^n$ and elliptic boundary value problems. *J. Funct. Anal.*, 114(2):286–347, 1993.

[7] B. Dahlberg. Weighted norm inequalities for the Lusin area integral and nontangential maximal functions for functions harmonic in a Lipschitz domain. *Studia Math.*, 67:297–314, 1980.

[8] B. Dahlberg and C. Kenig. Hardy spaces and the Neumann problem in L^p for Laplace's equation in Lipschitz domains. *Ann. of Math.*, 125:437–466, 1987.

[9] B. E. J. Dahlberg. Estimates of harmonic measure. *Arch. Rational Mech. Anal.*, 65(3):275–288, 1977.

[10] E. Fabes. Layer potential methods for boundary value problems on Lipschitz domains. In *Potential theory—surveys and problems*, volume 1344 of *Lecture notes in math.*, pages 55–80. Springer Verlag, 1988.

[11] D. Gilbarg and N. Trudinger. *Elliptic partial differential equations of second order.* Springer-Verlag, Berlin, 1983.

[12] P. Grisvard. *Elliptic problems in nonsmooth domains.* Pitman, 1985.

[13] D. Jerison and C. Kenig. The Neumann problem on Lipschitz domains. *Bull. Amer. Math. Soc.*, 4:203–207, 1982.

[14] C. E. Kenig. *Harmonic analysis techniques for second order elliptic boundary value problems.* Published for the Conference Board of the Mathematical Sciences, Washington, DC, 1994.

[15] G. M. Lieberman. Mixed boundary value problems for elliptic and parabolic differential equations of second order. *J. Math. Anal. Appl.*, 113(2):422–440, 1986.

[16] G. M. Lieberman. Optimal Hölder regularity for mixed boundary value problems. *J. Math. Anal. Appl.*, 143(2):572–586, 1989.

[17] J. Pipher and G. Verchota. The Dirichlet problem in L^p for the biharmonic equation on Lipschitz domains. *Amer. J. Math.*, 114(5):923–972, 1992.

[18] F. Rellich. Darstellung der eigenwerte von $\Delta u + \lambda u$ durch ein Randintegral. *Math. Z.*, 46:635–646, 1940.

[19] G. Savaré. Regularity and perturbation results for mixed second order elliptic problems. *Comm. Partial Diff. Eqns.*, 22:869–899, 1997.

[20] J. Serrin and H. F. Weinberger. Isolated singularities of solutions of linear elliptic equations. *Amer. J. Math.*, 88:258–272, 1966.

[21] E. Stein. *Singular integrals and differentiability properties of functions.* Princeton University Press, Princeton NJ, 1970.

[22] E. Stein and G. Weiss. On the theory of harmonic functions in several variables, I. *Acta. Math.*, 103:25–62, 1960.

[23] J. Sykes. *L^p regularity of solutions of the mixed boundary value problem for Laplace's equation on a Lipschitz graph domain.* PhD thesis, University of Kentucky, 1999.

[24] A. Torchinsky. *Real variable methods in harmonic analysis.* Academic Press, 1986.

[25] G. Verchota. Layer potentials and regularity for the Dirichlet problem for Laplace's equation on Lipschitz domains. *J. Funct. Anal.*, 59:572–611, 1984.

Department of Mathematics, University of Kentucky, Lexington, KY 40506
Current address: Department of Mathematics, West Virginia Wesleyan College, Buckhannon, WV 26201
E-mail address: sykes@wvwc.edu

Department of Mathematics, University of Kentucky, Lexington, KY 40506
E-mail address: rbrown@uky.edu
URL: http://www.ms.uky.edu/~rbrown

Contemporary Mathematics
Volume **277**, 2001

Sub-elliptic Besov spaces and the characterization of traces on lower dimensional manifolds

Donatella Danielli, Nicola Garofalo, and Duy-Minh Nhieu

1. General theory

In classical analysis Besov spaces play a fundamental role in the study of boundary value problems, see, e.g., [**LiMa**], [**Gris**], and also the recent papers [**JK**], [**FMM**]. One of their most basic aspect is that they serve to characterize the traces on the boundary of functions belonging to Sobolev spaces. In the paper [**DGN2**] we have studied this question in the setting of the Sobolev spaces relative to a system $X = \{X_1, ..., X_m\}$ of C^∞ vector fields in $\mathbb{R}^n$ satisfying the well-known finite rank condition due to Hörmander

$$rank\ Lie[X_1, ..., X_m] \ \equiv\ n.$$

Our main results can be summarized as follows. We introduce a suitable notion of Besov space with respect to a measure μ, and we prove that the elements of such space *are the traces* on the support of μ of the sub-elliptic Sobolev space associated with the system X. By this we mean that, under suitable assumptions on μ, and on the ground domain, we are able to construct appropriate trace and extension operators from the Sobolev to the Besov space, with the property that their composition is the identity. In this note we present the main results in [**DGN2**], but without going into the rather elaborate technical details. We mention that the theory outlined below constitutes a continuation (and also the completion) of the investigation in [**DGN1**], where we studied the question of the optimal Lebesgue spaces for the traces of Sobolev functions. One of the principal motivations behind our work was the study of boundary value problems for sub-Laplacians, with special

---

2000 *Mathematics Subject Classification.* Primary 46E35; Secondary 35H20.

Key words and phrases. Trace, restriction, extension, sub-elliptic Sobolev spaces, Besov spaces.

Second author was supported in part by NSF Grant No. DMS-0070492.

Third author was supported by NSC Grant No. 89-2115-M-001-018.

attention to the Neumann problem. For this aspect we refer the reader to the work in preparation [**DGN3**].

To state our main results we need to introduce the relevant notations and definitions. Given a system X as above, the first-order sub-elliptic Sobolev spaces are defined by

$$\mathcal{L}^{1,p}(\Omega, dx) \ = \ \{f \in L^p(\Omega, dx) \,|\, X_j f \in L^p(\Omega, dx), \ j = 1, ..., m\}, \qquad 1 \le p \le \infty,$$

where $X_j f$ denotes the weak derivative of f with respect to X_j. When endowed with the obvious norm, $\mathcal{L}^{1,p}(\Omega, dx)$ becomes a Banach space. Here, dx represents ordinary Lebesgue measure in $\mathbb{R}^n$. The notation $L^p(\Omega, dx)$ is not dictated by the desire of being pedantic, but rather by the need to keep a distinction with the Lebesgue spaces, to be introduced next, with respect to a different measure $d\mu$. The Lebesgue measure of a set $E \subset \mathbb{R}^n$ will always be indicated by $|E|$. We denote by $d(x,y)$ the Carnot-Carathéodory distance corresponding to the system X. For its definition and main properties we refer the reader to [**NSW**], or also [**G**]. It is well known that the metric balls $B(x,r) = \{y \in \mathbb{R}^n \mid d(y,x) < r\}$ satisfy the following basic property [**NSW**]: *For each bounded set $U \subset \mathbb{R}^n$, there exist constants $C_1, R_o > 0$ such that for any $x \in U$ and $0 < r \le R_o$ one has*

$$|B(x,2r)| \ \le \ C_1 \ |B(x,r)|. \tag{1.1}$$

Henceforth, we will refer to the numbers C_1 and R_o as the *characteristic local parameters* of U with respect to the system X. The number $Q = log_2 C_1$ is called the *local homogeneous dimension* associated with the system X and the set U.

We now introduce the relevant Besov spaces.

DEFINITION 1.1. Let μ be a non-negative Borel measure, having *supp* $\mu \subseteq F$, where F is a closed subset of $\mathbb{R}^n$. Given $1 \le p < \infty$, $0 < \beta < 1$, and $s > 0$ we define the semi-norm

$$\mathcal{N}^p_\beta(f, F, d\mu) \ = \ \left\{ \int_F \int_F \left(\frac{|f(x) \ - \ f(y)|}{d(x,y)^\beta} \right)^p \frac{d(x,y)^s}{|B(x, d(x,y))|} \, d\mu(y) \, d\mu(x) \right\}^{\frac{1}{p}}.$$

The *Besov space* on F, relative to the measure μ, is defined as

$$B^p_\beta(F, d\mu) \ = \ \{f \in L^p(F, d\mu) \,|\, \mathcal{N}^p_\beta(f, F, d\mu) < \infty\}.$$

If $f \in B^p_\beta(F, d\mu)$, the Besov norm of f is given by

$$\|f\|_{B^p_\beta(F,d\mu)} \ = \ \|f\|_{L^p(F,d\mu)} \ + \ \mathcal{N}^p_\beta(f, F, d\mu).$$

Some comments are in order. First, unlike what happens for Lebesgue measure, see (1.1), we do not assume μ to be a doubling measure. Secondly, the reader will have noticed that the parameter s, which is present in the right-hand side of the definition of the semi-norm, yet does not appear in the left-hand side $\mathcal{N}^p_\beta(f, F, d\mu)$. This omission is intentional, and is dictated by the desire of not overburdening the notation. Typically, s will be related to the measure μ by the assumption that the latter satisfy either one, or both, the following growth restrictions of Ahlfors type: *For every bounded set $U \subset \mathbb{R}^n$, with characteristic local parameters C_1, R_o, there exist $s \geq 0$, and $M = M(U) > 0$, such that for $x \in U$, $0 < r \leq R_o$ one has*

$$\mu(B(x,r)) \leq M \frac{|B(x,r)|}{r^s}, \tag{1.2}$$

or

$$\mu(B(x,r)) \geq M \frac{|B(x,r)|}{r^s}. \tag{1.3}$$

The specific assumptions on μ and s will be clearly spelled in the statement of each theorem. In the Euclidean setting, for instance, when $X = \{\partial/\partial x_1, ..., \partial/\partial x_n\}$, and μ is taken to be the $(n-1)$-dimensional Hausdorff measure restricted to the boundary of a Lipschitz domain, then $d(x,y) = |x-y|$, and (1.2) and (1.3) hold with $s = 1$. Thirdly, the parameter β in Definition 1.1 measures "smoothness". In this respect, it is interesting to observe that if (1.2) holds, and if $\beta' > \beta$, then the subelliptic Hölder class $\Gamma^{0,\beta'}(F)$ (for whose definition in the context of homogeneous groups we refer to [**FS**]) is continuously embedded into $B^p_\beta(F, d\mu)$.

It is important to note that, in most applications, the measure μ in Definition 1.1 is supported on a lower dimensional manifold, and therefore on a set of Lebesgue measure zero. Thereby, a function in L^p with respect to Lebesgue measure may not see the support of μ. To bypass this obstacle we introduce the following.

DEFINITION 1.2. For a function $f \in L^1_{loc}(\mathbb{R}^n, dx)$ we let

$$f^\star(x) = \begin{cases} \lim_{r\to 0} \frac{1}{|B(x,r)|} \int_{B(x,r)} f(y)\, dy & \text{if this limit exists,} \\ 0 & \text{otherwise.} \end{cases}$$

We call $f^\star$ the *precise representation* of f.

We observe that, thanks to (1.1), the Lebesgue differentiation theorem for a space of homogeneous type in [**C**] applies. From the latter we conclude that $f = f^\star$ $dx - a.e.$, but the equivalence of f and $f^\star$ with respect to a different measure μ is a delicate matter (and, in fact, such equivalence can not possibly subsist in general). The following basic result enables us to work with the precise representation of a Sobolev function f. It should be interpreted as a notion of trace in a qualitative sense.

THEOREM 1.3. *Let μ be a Borel measure satisfying* (1.2) *for $0 \le s < Q = Q(U)$. For any fixed bounded set $U \subset \mathbb{R}^n$, with local parameters C_1 and R_o, there exists $\delta > 1$ such that for $x_o \in U$, $0 < R < R_o/\delta$ and $f \in \mathcal{L}^{1,p}(B(x_o, \delta R), dx)$ one has*

$$\lim_{r\to 0} \frac{1}{|B(x,r)|} \int_{B(x,r)} |f(x) - f(y)|\, dy = 0 \qquad \text{for} \quad \mu - a.e. \quad x \in B(x_o, \delta R).$$

We stress that the conclusion of Theorem 1.3 is relative to the measure μ, and not to the Lebesgue measure dx. A different way of stating Theorem 1.3 is that $f(x) = f^\star(x)$, for μ-a.e. $x \in B(x_o, \delta R)$.

We can now state the first trace theorem for Sobolev functions with respect to Lebesgue measure. This is one of the most technical results in [**DGN2**]. Its proof rests on various delicate estimates whose discussion we do not attempt here, referring the reader to [**DGN2**]. In the sequel, if $B = B(x,r)$ is a ball, and $\sigma > 0$, we denote by σB the concentric ball $B(x, \sigma r)$.

THEOREM 1.4 (**Interior sharp trace inequality**). *Let $U \subset \mathbb{R}^n$ be a bounded set with characteristic local parameters C_1, R_o. There exists $\sigma = \sigma(X,U) > 0$ such that, given $p > 1$, $B_o = B(x_o, R) \subset U$, $0 < R < \frac{R_o}{2\sigma}$, $f \in \mathcal{L}^{1,p}(\sigma B_o, dx)$, and a non-negative Borel measure μ satisfying for some $0 < s < p$*

$$\mu(B(x,r)) \le M \frac{|B(x,r)|}{r^s}, \qquad x \in U,\ 0 < r \le R_o, \tag{1.4}$$

and such that supp $\mu = F \subset B_o$, and $\mu(F) > 0$, one has for any $0 < \beta \le 1 - s/p$

$$\|f\|_{B^p_\beta(F,d\mu)} \le C \|f\|_{\mathcal{L}^{1,p}(\sigma B_o, dx)}, \tag{1.5}$$

for some $C = C(X,U,p,s,\beta) > 0$.

Theorem 1.4 is of a local nature. To pass to a global version, we need to somehow extend a Sobolev function outside of its original domain. In order to do so, some geometric assumptions must be imposed on the ground domain. A very natural one is that contained in the following definition.

DEFINITION 1.5. An open set $\Omega \subset \mathbb{R}^n$ is called an (ϵ, δ)-domain if there exist $0 < \delta \le \infty$, $0 < \epsilon \le 1$, such that for any pair of points $p, q \in \Omega$, if $d(p,q) \le \delta$, then one can find a continuous, rectifiable curve $\gamma : [0,T] \to \Omega$, for which $\gamma(0) = p$, $\gamma(T) = q$, and

$$l(\gamma) \le \frac{1}{\epsilon}\, d(p,q), \qquad d(z, \partial\Omega) \ge \epsilon \, \min\{d(p,z), d(z,q)\} \quad \text{for all } z \in \{\gamma\}.$$

If $\Omega \subset \mathbb{R}^n$ is an open set, then *the radius* of Ω is defined as follows

$$rad(\Omega) \;=\; sup\,\{r > 0 \mid \partial B(p,s) \cap \Omega \neq \varnothing, \text{for every } p \in \Omega, \text{ and every } 0 \le s < r\}.$$

Note that if $\delta = \infty$, or Ω is connected, then $rad(\Omega) > 0$. We also remark that, thanks to Lebesgue differentiation theorem in a space of homogeneous type [**C**], the Lebesgue measure of the boundary of an (ϵ,δ)-domain Ω is zero, i.e.,

$$|\partial\Omega| \;=\; 0. \tag{1.6}$$

The next basic result established in [**GN**] plays an important role in the above mentioned passage from local to global.

THEOREM 1.6. *Let* $1 \le p \le \infty$. *If* $\Omega \subset \mathbb{R}^n$ *is a bounded* (ϵ,δ)*-domain with* $rad(\Omega) > 0$, *then there exists a linear operator* $\mathcal{S} : \mathcal{L}^{1,p}(\Omega, dx) \longrightarrow \mathcal{L}^{1,p}(\mathbb{R}^n, dx)$ *such that for some* $C > 0$ *one has for* $f \in \mathcal{L}^{1,p}(\Omega, dx)$

$$(i)\, \mathcal{S}f(x) \;=\; f(x) \quad \text{for a.e. } x \in \Omega, \qquad (ii)\, \|\mathcal{S}f\|_{\mathcal{L}^{1,p}(\mathbb{R}^n,dx)} \le C\|f\|_{\mathcal{L}^{1,p}(\Omega,dx)}.$$

We have used here $\mathcal{S}$ to denote the extension operator for Sobolev spaces, and will reserve the symbol $\mathcal{E}$ to indicate extensions from Besov spaces, to be introduce below.

Combining Theorem 1.4 with Theorem 1.6, we obtain.

THEOREM 1.7 (**Trace theorem on the boundary**). *Let* $U \subset \mathbb{R}^n$ *be a bounded set with characteristic local parameters* C_1, R_o, *and let* $p > 1$. *There is* $\sigma = \sigma(X,U) > 0$ *such that, if* $\Omega \subset U$ *is a bounded* (ϵ,δ)*-domain with* $rad(\Omega) > 0, diam(\Omega) < \frac{R_o}{2\sigma}$, $dist(\Omega, \partial U) > R_o$, *and* μ *is a positive Borel measure having supp* $\mu \subseteq \partial\Omega$, *and satisfying* (1.4) *for some* $0 < s < p$, *then for every* $0 < \beta \le 1 - s/p$ *there exist a linear operator*

$$\mathcal{T}r : \mathcal{L}^{1,p}(\Omega, dx) \;\rightarrow\; B^p_\beta(\partial\Omega, d\mu),$$

and a constant $C = C(U,p,s,\eta,\beta,\epsilon,\delta,rad(\Omega)) > 0$, *such that*

$$\|\mathcal{T}r\, f\|_{B^p_\beta(\partial\Omega,d\mu)} \;\le\; C\|f\|_{\mathcal{L}^{1,p}(\Omega,dx)}. \tag{1.7}$$

Furthermore, if $f \in C^\infty(\overline{\Omega}) \cap \mathcal{L}^{1,p}(\Omega, dx)$, *then* $\mathcal{T}r\, f = f$ *on* $\partial\Omega$.

Theorem 1.7 states that a function in the Sobolev space $\mathcal{L}^{1,p}(\Omega, dx)$ has a trace on the boundary of Ω which belongs to the Besov space $B^p_\beta(\partial\Omega, d\mu)$. However of crucial importance in the applications to boundary value problems, such result still falls short of a complete characterization of the elements of $B^p_\beta(\partial\Omega, d\mu)$ as traces of functions in $\mathcal{L}^{1,p}(\Omega, dx)$. This leads to introduce our next result.

THEOREM 1.8 (**Extension theorem**). *Let $1 \leq p < \infty$ and μ be a nonnegative Borel measure on $\mathbb{R}^n$ satisfying (1.3) with $0 < s < p$. When $p > n$ we request in addition that $s \leq \frac{n+p}{2}$. We assume that $F = supp\ \mu \subset \Omega$ be compact, with $|F| = 0$. There exists a bounded linear mapping (an* extension operator*) $\mathcal{E} : B^p_{1-\frac{s}{p}}(F, d\mu) \to \mathcal{L}^{1,p}(\Omega, dx)$, such that*

$$(i)\ \mathcal{E}f(x) = f(x) \quad \text{for } \mu \text{ a.e. } x \in F, \qquad (ii)\ \|\mathcal{E}f\|_{\mathcal{L}^{1,p}(\Omega,dx)} \leq C\|f\|_{B^p_{1-s/p}(F,d\mu)},$$

for some $C > 0$. Furthermore, $\mathcal{E}f$ is supported in a neighborhood of F.

As we will see subsequently, in the important situation when F is the boundary of a domain, the parameter s in (1.3) will be $s = 1$. Thus the limitation $0 < s < p$ is always satisfied if we confine attention to the range $p > 1$. Moreover, since $n \geq 3$ one also has $\frac{n+p}{2} > 2 > 1 = s$. Also, we would like to note that the limitation $s \leq \frac{n+p}{2}$ when $p > n$ in the statement of the theorem is a technical limitation in the case of Hörmander vector fields. When the system X generates the Lie algebra of a Carnot group $\boldsymbol{G}$ having homogeneous dimension Q, this constraint can be considerably improved. In such case, one has the much wider range $0 < s \leq \frac{Q+p}{2}$.

At this point we combine Theorems 1.7 and 1.8 to establish the central result of this note.

THEOREM 1.9. *Let $U \subset \mathbb{R}^n$ be a bounded set, with characteristic local parameters C_1, R_o, and $\Omega \subset U$ be an (ϵ, δ)-domain, with $rad(\Omega) > 0$, $dist(\Omega, \partial U) > R_o$. Given $p > 1$, let μ be a non-negative Borel measure, with $supp\ \mu \subseteq \partial\Omega$, and satisfying for some $M > 0$, and $0 < s < p$, the Ahlfors condition*

$$M^{-1}\,\frac{|B(x,r)|}{r^s} \leq \mu(B(x,r)) \leq M\,\frac{|B(x,r)|}{r^s} \tag{1.8}$$

for all $x \in U$, $0 < r < R_o$. When $p > n$ we assume in addition that $s \leq \frac{n+p}{2}$. There is $\sigma = \sigma(U, X) > 0$ such that, if $diam(\Omega) < \frac{R_o}{2\sigma}$, then there exist two linear operators

$$\mathcal{T}r : \mathcal{L}^{1,p}(\Omega, dx) \to B^p_{1-s/p}(\partial\Omega, d\mu) \qquad \mathcal{E} : B^p_{1-s/p}(\partial\Omega, d\mu) \to \mathcal{L}^{1,p}(\Omega, dx)$$

such that $\mathcal{T}r \circ \mathcal{E}$ is the identity map from $B^p_{1-s/p}(\partial\Omega, d\mu)$ into itself.

Proof. We begin by observing that, thanks to (1.6), we have $|\partial\Omega| = 0$, therefore if $F = supp\ \mu$, we obtain in particular $|F| = 0$. We can thus apply Theorem 1.8, which guarantees the existence of an extension operator $\mathcal{E}$. The existence of a trace operator $\mathcal{T}r$ follows directly from Theorem 1.7. This completes the proof. □

Theorem 1.9 gives a complete characterization of the traces of Sobolev functions on the boundary for the general class of (ϵ, δ) domains. Thanks to such result it is now possible to say that *the space* $B^p_{1-s/p}(\partial\Omega, d\mu)$ *is the trace space of* $\mathcal{L}^{1,p}(\Omega, dx)$ *on the boundary of* Ω. We will express this property as follows

$$\mathcal{L}^{1,p}(\Omega, dx)|_{\partial\Omega} \;=\; B^p_{1-s/p}(\partial\Omega, d\mu).$$

It is interesting to compare the trace Theorem 1.4 with that established in [**DGN1**]. In that paper, under the same assumption on the measure μ, we proved that for $1 < p < Q$ the embedding

$$\mathcal{L}^{1,p}(\sigma B, dx) \;\subset\; L^q(B, d\mu), \qquad q = p\frac{Q-s}{Q-p} > p, \tag{1.9}$$

is continuous. Such result is sharp in the scale of L^q spaces. Since the Besov spaces are the trace spaces for the subelliptic Sobolev spaces, it is natural to ask whether it is possible to continuously embed $B^p_\beta(\Omega, d\mu)$ into the Lebesgue space $L^q(\Omega, d\mu)$ in (1.9). This is indeed the case if μ is assumed to satisfy the bound from below.

THEOREM 1.10 (**Embedding of Besov spaces into Lebesgue spaces**). *Given a bounded set* $U \subset \mathbb{R}^n$ *and parameters* C_1, R_o, Q *as in* (1.1), *let* $\Omega \subset \overline{\Omega} \subset U$ *be an open set with diam* $\Omega < R_o/2$. *Let* $p \geq 1$, $0 < \beta < 1$. *Suppose* μ *is a non-negative Borel measure with supp* $\mu = F \subset \Omega$, *and satisfying* (1.3) *with*

$$0 < s \;\leq\; n + \beta p, \qquad s \;<\; Q - \beta p.$$

There exists a continuous embedding

$$B^p_\beta(\Omega, d\mu) \;\subset\; L^q(\Omega, d\mu), \qquad \text{where} \;\; q \;=\; p\,\frac{Q-s}{Q-s-\beta p},$$

and, in fact, for $f \in B^p_\beta(\Omega, d\mu)$ *one has*

$$\|f\|_{L^q(\Omega,d\mu)} \leq C\left\{\left(1 + \frac{diam(\Omega)^\beta}{\mu(\Omega)^{\beta/(Q-s)}}\right)\mathcal{N}^p_\beta(f, F, d\mu) + \frac{1}{\mu(\Omega)^{\beta/(Q-s)}}\;\|f\|_{L^p(\Omega,d\mu)}\right\},$$

where $C = C(\Omega, C_1, R_o, p, \beta, M, s) > 0$. *Furthermore,*

$$\left(\int_\Omega |f(x) - f_{\Omega,\mu}|^q\, d\mu(x)\right)^{\frac{1}{q}} \leq C\left(\int_\Omega\int_\Omega |f(x) - f(y)|^p \frac{d(x,y)^{s-\beta\, p}}{|B(x, d(x,y))|}\, d\mu(y)\, d\mu(x)\right)^{\frac{1}{p}},$$

where $f_{\Omega,\mu}$ *denotes the average* $\frac{1}{\mu(\Omega)}\int_\Omega f\, d\mu$.

2. Geometric examples in Carnot groups

In this section we present some important examples of domains and measures μ satisfying conditions (1.2), (1.3), and (1.8). The geometric setting will be that of Carnot groups. In this context we will state various cases of the above theorems which are particularly relevant in the applications, see [**DGN3**]. One very important aspect, is that, thanks to the fact that in a Carnot group $\boldsymbol{G}$ the Nagel-Stein-Wainger polynomial is actually a monomial (see (2.6)), in (1.1) we can take

$U = \boldsymbol{G}$, and $R_o = \infty$. As a consequence, *all limitations on the diameter of the ground domain Ω in the results of section 1 can be dropped.* In the discussion of the examples below, we will omit all the relevant calculations, since most of them are rather technically involved. The reader is referred to [**DGN2**] for full details.

A Carnot group $\boldsymbol{G}$ of step r is a simply connected Lie group whose Lie algebra $\mathfrak{g}$ admits a nilpotent stratification up to step r. This means that $\mathfrak{g} = V_1 \oplus V_2 \oplus \cdots \oplus V_r$, and that moreover $[V_1, V_j] = V_{j+1}$ for $j = 1, ..., r-1$, whereas $[V_1, V_r] = \{0\}$. We assume that a scalar product $< \cdot, \cdot >$ is given on $\mathfrak{g}$ for which the $V_j's$ are mutually orthogonal. Denote with $m = dim\ V_1$, $k = dim\ V_2$. Throughout this section we indicate by $X = \{X_1, \ldots, X_m\}$ a fixed orthonormal basis of the first layer V_1, and continue to denote by X the corresponding system of left-invariant sections on $\boldsymbol{G}$. Also, $Y_1, ..., Y_k$ will denote a fixed orthonormal basis of the second layer V_2. The exponential map $exp : \mathfrak{g} \to \boldsymbol{G}$ is a global analytic diffeomorphism. It allows to define analytic maps $\xi_i : \boldsymbol{G} \to V_i$, $i = 1, ..., r$, by letting $g = \exp(\xi_1(g) + ... + \xi_r(g))$. We set

(2.1)
$$x_j(g) \ = < \xi_1(g), X_j >, \quad j = 1, ..., m, \qquad y_s(g) \ = < \xi_2(g), Y_s >, \quad s = 1, ..., k.$$

If we let $x(g) = (x_1(g), ..., x_m(g))$, $y(g) = (y_1(g), ..., y_k(g))$, then $x(g), y(g)$ respectively indicate the projection onto the first and second layers V_1, and V_2, of the exponential coordinates of $g \in \boldsymbol{G}$. When $\boldsymbol{G}$ is of step 2, then the exponential coordinates of $g \in \boldsymbol{G}$ are $(x(g), y(g))$.

Let $N = dim(V_1) + \cdots + dim(V_r)$ indicate the topological dimension of $\boldsymbol{G}$. We use letters g_o, g, g' for points in $\boldsymbol{G}$, and denote by dg the bi-invariant Haar measure on $\boldsymbol{G}$ obtained by lifting via the exponential map exp the Lebesgue measure on $\mathfrak{g}$. Every Carnot group is naturally equipped with a family of non-isotropic dilations defined by

$$\delta_\lambda(g) \ = \ exp \circ \Delta_\lambda \circ exp^{-1}(g), \quad g \in \boldsymbol{G}, \tag{2.2}$$

where $\Delta_\lambda : \mathfrak{g} \to \mathfrak{g}$ is defined by $\Delta_\lambda(X_1 + \cdots + X_r) = \lambda X_1 + \cdots + \lambda^r X_r$. One easily checks that $(d \circ \delta_\lambda)(g) = \lambda^Q\ dg$, where $Q = \sum_{j=1}^r j\ dim(V_j)$ is the *homogeneous dimension* of G.

We denote by $d(g, g')$ the Carnot-Carathéodory distance on $\boldsymbol{G}$ associated with the system X. It is well-known that $d(g, g')$ is equivalent to the gauge pseudo-metric $\rho(g, g')$ on $\boldsymbol{G}$ defined as follows. Let $|\cdot|$ denote the Euclidean distance to the origin on $\mathfrak{g}$. For $\xi = \xi_1 + \cdots + \xi_r \in \mathfrak{g}$, $\xi_i \in V_i$, one lets

$$|\xi|_{\mathfrak{g}} \ = \ \left(\sum_{i=1}^r |\xi_i|^{2r!/i} \right)^{2r!}, \qquad |g|_G \ = \ |\exp^{-1} g|_{\mathfrak{g}}, \qquad g \in \boldsymbol{G}. \tag{2.3}$$

The pseudo-distance on $\boldsymbol{G}$ associated to $|\cdot|_{\boldsymbol{G}}$ is given by

$$\rho(g,g') = |g^{-1}\, g'|_{\boldsymbol{G}}. \tag{2.4}$$

It is easy to see that there exists a constant $C = C(\boldsymbol{G}) > 0$ such that

$$C\,\rho(g,g') \le d(g,g') \le C^{-1}\,\rho(g,g'), \qquad g, g' \in \boldsymbol{G}, \tag{2.5}$$

see [**NSW**]. Denoting with $B(g,r) = \{g' \in \boldsymbol{G} \mid d(g',g) < r\}$ and $B_\rho(g,r) = \{g' \in \boldsymbol{G} \mid \rho(g',g) < r\}$, respectively the ball and pseudo-ball centered at g with radius r, one easily recognizes that there exist $\omega = \omega(\boldsymbol{G}) > 0$, and $\alpha = \alpha(\boldsymbol{G}) > 0$ such that

$$|B(g,R)| = \omega\, R^Q, \qquad |B_\rho(g,R)| = \alpha\, R^Q, \qquad g \in \boldsymbol{G}, R > 0. \tag{2.6}$$

The first equation in (2.6) shows, in particular, that for a Carnot group the Nagel-Stein-Wainger polynomial is simply the monomial ωR^Q.

The *sub-Laplacian* associated with the system X is the left-invariant second-order partial differential operator on $\boldsymbol{G}$ given by

$$\mathcal{L} = -\sum_{j=1}^{m} X_j^* X_j = \sum_{j=1}^{m} X_j^2 \tag{2.7}$$

By the assumption on the Lie algebra, thanks to Hörmander's theorem the operator $\mathcal{L}$ is hypoelliptic, although it fails to be elliptic when $r > 1$. For a function u on $\boldsymbol{G}$ we let $|Xu| = (\sum_{j=1}^m (X_j u)^2)^{1/2}$.

Consider the $(N-1)-$dimensional Hausdorff measure H_{N-1} constructed with the Riemannian distance of $\boldsymbol{G}$. Given a C^1 domain $\Omega \subset \boldsymbol{G}$ we will denote with $d\sigma$ the surface measure on $\partial\Omega$ defined by $\sigma = H_{N-1}\lfloor\partial\Omega$. (By abuse of notation, we also denote the surface measure on $\mathfrak{g}$ by $d\sigma = d\sigma \circ exp$). The *characteristic set* of Ω with respect to the system X is defined as follows

$$\Sigma = \Sigma_{\Omega,X} = \{g \in \partial\Omega \mid X_j(g) \in T_g(\partial\Omega), j = 1, ..., m\}.$$

Suppose that $\phi : \boldsymbol{G} \to \mathbb{R}$ be a C^1 defining function for Ω, i.e., $\Omega = \{g \in \boldsymbol{G} \mid \phi(g) < 0\}$. Denoting with ∇ the Riemannian gradient in $\boldsymbol{G}$, we always assume that $|\nabla\phi(g)| > 0$, for every $g \in \partial\Omega$. In particular, when Ω is a bounded set we infer the existence of a constant $\alpha_\Omega > 0$ such that

$$0 < \alpha_\Omega \le |\nabla\, \phi(g)|, \qquad \text{for every} \quad g \in \partial\Omega.$$

The *angle function* of Ω is defined by $w(g) \overset{def}{=} |X\phi(g)|$. Since the outer unit normal to $\partial\Omega$ is given by $\nu = \nabla\phi/|\nabla\phi|$, it is clear that

$$\Sigma = \{g \in \partial\Omega \mid w(g) = 0\}.$$

Typically, bounded domains have non-empty characteristic sets. For instance, due to topological reasons every bounded C^1 domain in the Heisenberg group $\mathbb{H}^n$, whose boundary is homeomorphic to the $2n$-dimensional sphere $\mathbb{S}^{2n}$, has non-empty characteristic set. The following result, which was proved by Derridj for general Hörmander type vector fields [**De1**], [**De2**], shows that, at least from the measure theoretic point of view, the set Σ is not too big.

THEOREM 2.1. *Let $\Omega \subset \boldsymbol{G}$ be a C^∞ domain. One has*

$$H_{N-1}(\Sigma_{\Omega,X}) \;=\; 0.$$

We next turn to the main objective of this section, i.e., presenting examples of measures μ satisfying the two basic hypothesis (1.2) and/or (1.3). We will in fact be solely concerned with the geometrically relevant case $s = 1$.

For a given C^1 open set $\Omega \subset \boldsymbol{G}$, with defining function ϕ, we introduce the following Borel measure in $\boldsymbol{G}$

$$\mu(E) \overset{def}{=} \int_{E\cap\partial\Omega} |X\phi(g)|\, d\sigma, \qquad E \quad \text{is a Borel subset of} \quad \boldsymbol{G}. \tag{2.8}$$

The measure μ is supported on $\partial\Omega$. It is important to observe that, when E is a C^1 domain, with defining function ϕ satisfying

$$0 \;<\; \alpha \;\le\; |\nabla\phi(g)|^{-1} \;\le\; \beta, \qquad\qquad g \in \partial\Omega,$$

then for every open set $\Omega \subset \boldsymbol{G}$ one has

$$\alpha\, P_X(\Omega, E) \;\le\; \mu(E) \;\le\; \beta P_X(\Omega, E). \tag{2.9}$$

In (2.9) we have denoted with $P_X(\Omega, E)$ the relative $X-$perimeter of Ω in E introduced in [**CDG**], and also independently in [**FGW**]. In the sequel, we will adopt the notation

$$\Delta(g, r) \;=\; B(g, r) \,\cap\, \partial\Omega, \qquad\qquad g \in \partial\Omega.$$

THEOREM 2.2. *Let $\boldsymbol{G}$ be a Carnot group of arbitrary step, and consider a C^2 domain*

$$\Omega \;=\; \{g \in \boldsymbol{G} \mid \phi(g) < 0\},$$

where $\phi \in C^2(\boldsymbol{G})$ is a defining function for Ω. There exist $M, R_o > 0$, depending on $\boldsymbol{G}$ and Ω, such that for every $g_o \in \partial\Omega$ and $0 < r \le R_o$ one has

$$\left(\sup_{\Delta(g_o,r)} |X\phi|\right) \sigma(\Delta(g_o, r)) \;\le\; M\, \frac{|B(g_o, r)|}{r}.$$

We stress that the estimate in Theorem 2.2 holds irregardless of whether the point g_o is characteristic or not. In the special setting of the Heisenberg group $\mathbb{H}^n$, Theorem 2.2 was first formulated independently by M. Mekias and by C. Romero, in their respective Ph.D. Dissertations [**Me**] and [**Ro**] (however, neither of these works is available in print, and also the proofs were not complete). For a detailed proof for the Heisenberg group the reader should see [**DGN1**]. Modulo simple modifications, such proof extends in fact to all Carnot groups of step two, see [**CGN1**]. In its general formulation above, Theorem 2.2 was established in [**CGN2**]. In the same paper, using the Rothschild-Stein lifting theorem [**RS**], a general version of Theorem 2.2 was also given.

We observe explicitly that the geometric estimate in Theorem 2.2 provides a precise quantitative information on how bad surface measure $d\sigma$ can be at characteristic points. For instance, in the case of a Carnot group of step two it was shown in [**DGN1**] that one has generically

$$\sigma(\Delta(g_o,r)) \leq M \frac{|B(g_o,r)|}{r^2},$$

and that such estimate is sharp when g_o is a characteristic point, in the sense that it cannot be improved to

$$\sigma(\Delta(g_o,r)) \leq M \frac{|B(g_o,r)|}{r^s},$$

for some $1 \leq s < 2$. These observations naturally lead to the question of what is the appropriate replacement for surface measure in the trace inequalities. The answer is provided by the measure μ defined by (2.8). To see why, we state the following important consequence of Theorem 2.2.

THEOREM 2.3. *Let $\boldsymbol{G}$ and Ω be as in Theorem 2.2. There exist $M, R_o > 0$, depending on $\boldsymbol{G}$ and Ω, such that for every $g_o \in \partial\Omega$, and $0 < r \leq R_o$, one has*

$$\mu(B(g_o,r)) \leq M \frac{|B(g_o,r)|}{r},$$

where μ is the measure, concentrated on $\partial\Omega$, introduced in (2.8).

Theorem 2.3 provides a basic class of geometric examples for Theorems 1.4, 1.7. We mention that in their recent interesting paper [**MM**], Monti and Morbidelli have proved a localized version of the estimate in Theorem 2.3 for general systems of Hörmander type. However, they need to assume that the relevant portion of the boundary be non-characteristic. Under this restriction the authors also prove that μ satisfies the estimate from below (1.3) with $s = 1$, and they use these results to obtain a local version, at non-characteristic points, of the trace Theorem 1.7.

We next state a particular case of Theorem 1.7 which is especially important in the applications.

THEOREM 2.4. *Let* $\boldsymbol{G}$ *be a Carnot group of arbitrary step* r, *and consider a* C^2, *bounded* (ϵ,δ)*-domain* $\Omega \subset \boldsymbol{G}$, *with* $rad(\Omega) > 0$. *Given* $p > 1$, *there exists a bounded linear operator* $\mathcal{T}r$, *called the* trace operator, *such that*

$$\mathcal{T}r : \mathcal{L}^{1,p}(\Omega, dg) \ \rightarrow \ B^p_{1-\frac{1}{p}}(\partial\Omega, d\mu),$$

where μ *is defined by* (2.8). *Moreover, one has* $\mathcal{T}rf = f$ *on* $\partial\Omega$, *when* $f \in C^\infty(\overline{\Omega})$.

To state the next result we need to introduce a definition. Recall the definition (2.1) of the exponential coordinates.

DEFINITION 2.5. Let $\boldsymbol{G}$ be a Carnot group of step 2 and consider a $C^{1,1}$ domain $\Omega \subset \boldsymbol{G}$. We say that Ω has *cylindrical symmetry near* Σ (in short, Ω is CS_Σ) if for each $g_o \in \Sigma$, after a group translation that sends g_o to the identity $e = (0,0)$, one can find a neighborhood U of e such that for some $l = 1, ..., k$ one has

$$\partial(g_o^{-1}\Omega) \ \cap \ U \ = \ \{g \in \boldsymbol{G} \mid y_l(g) = f(|x(g)|^4 + |y'(g)|^2)\} \cap U,$$

for some function $f \in C^{1,1}(\mathbb{R})$, with $f(0) = 0$.

The notation $y'(g)$ indicates the $(k-1)$- dimensional vector which is obtained from $y(g)$ after removing the l-th component $y_l(g)$. It is obvious that the gauge balls in $\boldsymbol{G}$ are CS_Σ domains. More in general, for $R > 0$ and $0 \le \lambda < R$, the sets

$$\Omega_{R,\lambda} \ = \ \{g \in \boldsymbol{G} \mid ((\lambda^2 + |v(g)|^2)^2 + |z(g)|^2)^{1/4} < R\}$$

are CS_Σ domains. By invariance under left-translation we conclude that also the sets $\Omega_{R,\lambda}(g_o) = g_o\Omega_{R,\lambda}$ are CS_Σ domains. The following basic result was established in [**CG**].

THEOREM 2.6. *In a Carnot group of step 2, every* CS_Σ *domain is a non-tangentially accessible domain (in short, NTA), with respect to the Carnot-Carathéodory metric* $d(g,g')$. *In particular, every* CS_Σ *domain is an* (ϵ,δ)*-domain.*

Because of its relevance, we state explicitly the following consequence of Theorem 2.6.

COROLLARY 2.7. Let $\boldsymbol{G}$ be a Carnot group of step 2. For every $g_o \in \boldsymbol{G}$, $R > 0$, and $0 \le \lambda < R$, the sets $\Omega_{R,\lambda}(g_o)$ are NTA, and therefore $(\epsilon,\delta)-$domains. In particular, the gauge balls are NTA domains.

If we apply Theorems 2.4 and 2.6 we obtain.

THEOREM 2.8. *Let $\boldsymbol{G}$ be a Carnot group of step 2, and consider a bounded CS_Σ-domain $\Omega \subset \boldsymbol{G}$ of class C^2. Given $p > 1$, there exists a bounded trace operator*

$$\mathcal{T}r : \mathcal{L}^{1,p}(\Omega, dg) \;\to\; B^p_{1-\frac{1}{p}}(\partial\Omega, d\mu),$$

where μ is defined by (2.8)*. Moreover, one has $\mathcal{T}rf = f$ on $\partial\Omega$, when $f \in C^\infty(\overline{\Omega})$. This holds in particular when Ω is a gauge ball.*

We now turn to the discussion of the bound from below in (1.3). Due to the presence of characteristic points on the boundary, the construction of examples of measures satisfying such an estimate is much more delicate than the corresponding one for (1.2). Eventually, our goal is to provide a large class of examples of measures satisfying both such assumptions. We begin with a definition.

DEFINITION 2.9. Given a C^1 open set $\Omega \subset \boldsymbol{G}$, with defining function ϕ, we say that its characteristic set Σ is *controlled* if there exists a constant $C > 0$, and a neighborhood U of Σ, such that for every $g \in U$ one has

$$w(g) \;=\; |X\phi(g)| \;\geq\; C\, d(g, \Sigma). \tag{2.10}$$

In (2.10) the distance can be measured with respect to the intrinsic distance $d(g, g')$ or, equivalently, with respect to the gauge pseudo-metric $\rho(g, g')$. Our main result concerning the estimate from below is the following.

THEOREM 2.10. *Let $\boldsymbol{G}$ be a Carnot group of step 2, and consider a bounded C^2 domain $\Omega \subset \boldsymbol{G}$. There exist positive constants M, R_o, depending on Ω, and on $\boldsymbol{G}$, such that if either $\Sigma = \Sigma_{\Omega,X} = \varnothing$, or Σ is controlled, then for every $g_o \in \partial\Omega$, and any $0 < r < R_o$, one has*

$$\mu(B(g_o, r)) \;\geq\; M^{-1}\, \frac{|B(g_o, r)|}{r}.$$

In some situations, it is possible that the relevant domain does not have characteristic points, and thereby (2.10) holds trivially. Here are two interesting examples.

EXAMPLE 2.11 (**Non-characteristic hyperplanes in Carnot groups**). Let $\boldsymbol{G}$ be a Carnot group. For a fixed vector $a \in \mathbb{R}^m \setminus \{0\}$ (recall that $m = dim\ V_1$), and for $\lambda \in \mathbb{R}$, we define the *half-space*

$$\mathcal{H}_a^+ \;=\; \{g \in \boldsymbol{G} \,|< x(g), a >> \lambda\}.$$

We call the set $H_a = \{g \in \boldsymbol{G} \mid < x(g), a >= \lambda\} = \partial\mathcal{H}_a^+$ a *non-characteristic hyperplane* in $\boldsymbol{G}$. The reason for the name is in the fact that one has $\Sigma = \Sigma_{H_a^+,X} = \varnothing$. In fact, the relative angle function satisfies the equation $w(g) \equiv |a|$, for every $g \in \boldsymbol{G}$.

A much less obvious example of a non-characteristic compact manifold is described as follows. Consider a Carnot group of step two $\boldsymbol{G}$, with Lie algebra $\mathfrak{g} = V_1 \oplus V_2$. Let M be a k-dimensional compact submanifold, with $\partial M = \varnothing$, of the $(k+1)$-dimensional vector space

$$\{\xi \in V_1 \mid < \xi, X_2 >= ... =< \xi, X_m >= 0\} \times V_2.$$

We assume in addition that

$$d_e(M, \{0\} \times V_2) \ > \ 0,$$

where $d_e(M, \{0\} \times V_2)$ denotes the Euclidean distance in $V_1 \times V_2$. If $\mathbb{S}^{m-1}$ denotes the $(m-1)-$dimensional sphere in V_1 centered at the origin, we define the *toroid* generated by M as

$$\mathbb{T}(M) \ = \ \{g \in \boldsymbol{G} \mid \exp^{-1}(g) \in \mathbb{S}^{m-1} \ \times \ M\}.$$

One has the following result.

EXAMPLE 2.12 (**A toroid in a group of Heisenberg type $\boldsymbol{G}$**). Let $\mathbb{T}(M) \subset \boldsymbol{G}$ be a toroid in group of Heisenberg type $\boldsymbol{G}$, then

$$\Sigma \ = \ \Sigma_{\mathbb{T}(M),X} \ = \ \varnothing.$$

For the reader's convenience we recall the definition of group of Heisenberg type introduced by A. Kaplan [**K**]. Let $\boldsymbol{G}$ be a group of step 2, with Lie algebra $\mathfrak{g} = V_1 \oplus V_2$. Consider consider the map $J : V_2 \to End(V_1)$ defined by

$$< J(\eta)\xi', \xi'' > \, = \, < [\xi', \xi''], \eta >, \qquad \eta \in V_2, \quad \xi', \xi'' \in V_1.$$

One says that $\boldsymbol{G}$ is of *Heisenberg type* if for every $\eta \in V_2$, such that $|\eta| = 1$, the map $J(\eta) : V_1 \to V_1$ is orthogonal. When the center V_2 of the group is one-dimensional, then there is essentially one such group, the Heisenberg group $\mathbb{H}^n$. The situation is, however, quite different when $dim\ V_2 > 1$, and it is possible to show that there is a plentiful supply of groups of Heisenberg type. For instance, the nilpotent component N in the Iwasawa decomposition $\boldsymbol{G} = KAN$ of a simple group of rank one, is a group of Heisenberg type. We refer the reader to [**DGN2**] for a detailed discussion.

The previous two examples provide instances in which the characteristic set is empty, and thereby trivially controlled. There exist in fact large classes of domains whose characteristic set is non-empty, but nonetheless controlled. A very basic one is the following.

PROPOSITION 2.13. In a group of Heisenberg type $\boldsymbol{G}$ the characteristic set of a gauge ball

$$B(g_o, r) \;=\; \{g' \in \boldsymbol{G} \mid \rho(g_o, g') < r\}$$

is controlled.

We next list other examples which, for simplicity, are stated in the setting of the first Heisenberg group $\mathbb{H} = \mathbb{R}^3$. We need the generators of the Lie algebra

$$X \;=\; \frac{\partial}{\partial x} + 2y\frac{\partial}{\partial t}, \qquad\qquad Y \;=\; \frac{\partial}{\partial y} - 2x\frac{\partial}{\partial t},$$

for which one has $[X, Y] = -4\partial/\partial t$. The gauge-distance is $d(g, g_o) = N(g^{-1}g_o)$, where

$$N(g) \;=\; N(x, y, t) \;=\; [(|x|^2 + |y|^2)^2 + t^2]^{\frac{1}{4}}.$$

1. Let $\mathcal{M} \subset \mathbb{H}^1$ be a manifold whose defining function is given by

$$\phi(x, y, t) \;=\; t \;-\; f(|z|^4)$$

where $f : [0, \infty) \to \mathbb{R}$ is a C^2 function and $z = (x, y)$, $|z| = \sqrt{x^2 + y^2}$. Without loss of generality, we may assume $f(0) = 0$. We easily see that

$$X\rho(z, t) \;=\; 4y \;-\; 8x|z|^2 f'(|z|^4), \quad Y\rho(z, t) \;=\; -4x \;-\; 8y|z|^2 f'(|z|^4)$$

and

$$w(z, t)^2 \;=\; X\rho(z, t)^2 \;+\; Y\rho(z, t)^2 \;=\; 16|z|^2(1 \;+\; 4|z|^4 f'(|z|^4)^2).$$

Hence, $\Sigma = \{(0, 0, 0)\}$ and $d((z, t), \Sigma)^4 = |z|^4 + f(|z|^4)^2$. Since f is a C^2 function and $f(0) = 0$, it is easy to see that

$$\lim_{|z| \to 0} \frac{w(z, t)^4}{d((z, t), \Sigma)^4} = \lim_{|z| \to 0} \frac{4|z|^4(1 \;+\; 4|z|^4 f'(|z|^4)^2)^2}{|z|^4(1 \;+\; |z|^{-4} f(|z|^4)^2)} = 4 \lim_{\tau \to 0} \frac{1 \;+\; 4\tau f'(\tau)^2}{1 \;+\; \tau^{-1} f(\tau)^2} > 0.$$

Thus, (2.10) follows. The same type of argument also applies to manifolds $\mathcal{M}$ whose defining function takes on the form

$$\phi(z, t) \;=\; t^2 \;-\; f(|z|^4)^2.$$

2. A different situation is provided by *the saddle*, given by

$$\phi(x, y, t) \;=\; t \;-\; x^2 \;+\; y^2.$$

It is easy to see that $w(x,y,t) = 2\sqrt{2}|x-y|$, and therefore $\Sigma = \{(x,y,0) | x = y\}$. We stress that the characteristic points of the saddle *are not* strongly isolated, according to the definition in [**J**]. Yet, the characteristic set Σ is controlled. In fact, using calculus, one can easily recognize that $d((x,y,x^2-y^2),\Sigma) = \frac{1}{\sqrt{2}}|x-y|$ and hence, w satisfies (2.10).

With Theorems 2.3 and 2.10 in hand, we can now state an important specific result.

THEOREM 2.14. *Let $\boldsymbol{G}$ be a Carnot group of step 2, and consider a C^2 domain $\Omega \subset \boldsymbol{G}$ having empty, or controlled characteristic set. There exist positive constants M, R_o, depending on Ω, and on $\boldsymbol{G}$, such that for every $g_o \in \partial\Omega$, and any $0 < r < R_o$, one has*

$$M^{-1}\,\frac{|B(g_o,r)|}{r} \;\le\; \mu(B(g_o,r)) \;\le\; M\,\frac{|B(g_o,r)|}{r}.$$

Here, μ is the measure defined by (2.8). In particular, this result holds when Ω is a gauge ball in a group of Heisenberg type.

Using Theorems 1.9, 2.6 and 2.14 we obtain the following result which plays a crucial role in the applications.

THEOREM 2.15. *Let $B(g_o,r)$ be a gauge ball in a group of Heisenberg type $\boldsymbol{G}$, having homogeneous dimension Q. For any $1 < p < Q$, and every $0 < \beta \le 1 - s/p$, there exist two linear operators*

$$\mathcal{T}r : \mathcal{L}^{1,p}(B(g_o,r),dg) \;\to\; B^p_{1-s/p}(\partial B(g_o,r),d\mu),$$

and

$$\mathcal{E} : B^p_{1-s/p}(\partial B(g_o,r),d\mu) \;\to\; \mathcal{L}^{1,p}(B(g_o,r),dg),$$

such that $\mathcal{T}r \circ \mathcal{E}$ is the identity map from $B^p_{1-s/p}(\partial\Omega,d\mu)$ into itself.

3. The Neumann problem

We close this note by discussing a notable application of the results which we have presented in the previous sections: *the subelliptic Neumann problem.* Let $X = \{X_1,...,X_m\}$ be a system of smooth vector fields of Hörmander type in $\mathbb{R}^n$. We consider the second order partial differential operator $\mathcal{L} = \sum_{j=1}^m X_j^* X_j$, where

X_j^* denotes the formal adjoint of X_j in L^2. Given a sufficiently nice (say, C^∞) domain $\Omega \subset \mathbb{R}^n$, we consider the problem

$$\begin{cases} \mathcal{L}u = 0 & \text{in } \Omega, \\ \sum_{j=1}^m X_j u < X_j, \eta > = \nu & \text{on } \partial\Omega \end{cases} \tag{3.1}$$

where η is the unit outer normal to Ω. Here, $\nu \in B^2_{\frac{1}{2}}(\partial\Omega, d\mu)^*$, with

$$< \nu, 1 > = 0. \tag{3.2}$$

The notation $B^2_{\frac{1}{2}}(\partial\Omega, d\mu)^*$ indicates the dual of the Besov space $B^2_{\frac{1}{2}}(\partial\Omega, d\mu)$, and the symbol $< \cdot, \cdot >$ represents the duality between these spaces. When $\nu = g \in L^1(\partial\Omega, d\sigma)$, then (3.2) reads

$$\int_{\partial\Omega} g \, d\sigma = 0.$$

In order to even introduce the notion of variational solution to (3.1), one needs to have a result which guarantees the existence, in an appropriate sense, of the traces of Sobolev functions on the boundary of Ω. This is where our results, and specifically Theorem 1.7, come into the play. We assume henceforth that Ω is a (ϵ, δ)-domain, with $rad(\Omega) > 0$, and that μ is a non-negative Borel measure supported in $\partial\Omega$, and satisfying the estimate (1.2). According to Theorem 1.7, there exists a continuous trace operator $Tr : \mathcal{L}^{1,2}(\Omega, dx) \to B^2_{\frac{1}{2}}(\partial\Omega, d\mu)$ such that

$$Tr(\phi) = \phi|_{\partial\Omega},$$

when $C^\infty(\bar{\Omega})$. We next introduce the Hilbert space

$$\tilde{\mathcal{L}}^{1,2}(\Omega, dx) = \{f \in \mathcal{L}^{1,2}(\Omega, dx) \,|\, \frac{1}{|\Omega|} \int_\Omega f(x)\, dx = 0\},$$

endowed with the obvious norm. By a *variational solution* to the problem (3.1) we mean a function $u \in \mathcal{L}^{1,2}(\Omega, dx)$ such that

$$\int_\Omega < Xu, X\phi > dx = - < \nu, Tr(\phi) > \tag{3.3}$$

for any $\phi \in \tilde{\mathcal{L}}^{1,2}(\Omega)$.

The following theorem holds.

THEOREM 3.1. *If ν satisfies the compatibility condition* (3.2), *the Neumann problem* (3.1) *admits a unique variational solution.*

For instance, a basic situation to which Theorem 3.1 applies is the one depicted in Theorem 2.15. Having established the existence of a variational solution u, one is interested in studying its fine properties, i.e., a priori estimates for the maximal function of the sub-gradient Xu, non-tangential behavior, etc. This is the subject of [**DGN3**].

Acknowledgement: The third author would like to thank the organizers of the University of Arkansas' Spring Lecture Series 2000, and Luca Capogna in particular, for the invitation to deliver in March 2000 the lecture to which this note is closely related.

References

[C] A.-P. Calderón, *Inequalities for the maximal function relative to a metric*, Studia Math., **57** (1976), 3, 297-306.

[CDG] L. Capogna, D. Danielli & N. Garofalo, *The geometric Sobolev embedding for vector fields and the isoperimetric inequality*, Comm. Anal. and Geom., **2** (1994), 201-215.

[CG] L. Capogna & N. Garofalo, *Boundary behavior of non-negative solutions of subelliptic equations in NTA domains for Carnot-Carathéodory metrics*, Journal of Fourier Anal. and Appl.,4-5, **4** (1998), 403-432.

[CGN1] L. Capogna, N. Garofalo & D. M. Nhieu, *Properties of harmonic measures in the Dirichlet problem for nilpotent Lie groups of Heisenberg type*, preprint (1999).

[CGN2] ———, *The Dirichlet problem for operators of Hörmander type*, preprint (1998).

[DGN1] D. Danielli, N. Garofalo & D. M. Nhieu, *Trace inequalities for Carnot-Carathéodory spaces and applications*, Ann. Sc. Norm. Sup. Pisa, Cl. Sci. (4), 2, **27** (1998), 195-252.

[DGN2] ———, *The trace space of Sobolev functions for Hörmander vector fields*, preprint, (1999).

[DGN3] ———, *The subelliptic Neumann problem*, in preparation.

[De1] M. Derridj , *Un probléme aux limites pour une classe d'opérateurs du second ordre hypoelliptiques*, Ann. Inst. Fourier, Grenoble, **21**, 4 (1971), 99-148.

[De2] ———, *Sur un théorème de traces*, Ann. Inst. Fourier, Grenoble, **22**, 2 (1972), 73-83.

[FMM] E. Fabes, O. Mendez & M. Mitrea, *Boundary layers on Sobolev-Besov spaces and Poisson's equation for the Laplacian in Lipschitz domains*, J. Funct. Anal., **159** (1998), 323-368.

[FS] G.B. Folland & E.M. Stein, *Hardy Spaces on Homogeneous Groups*, Princeton Univ. Press., (1982).

[FGW] B. Franchi, S. Gallot & R. Wheeden, *Sobolev and isoperimetric inequalities for degenerate metrics*, Math. Ann., **300** (1994), 557-571.

[G] N. Garofalo, *Recent Developments in the Theory of Subelliptic Equations and Its Geometric Aspects*, Birkhäuser, book in preparation.

[GN] N Garofalo & D.M. Nhieu, *Lipschitz continuity, global smooth approximations and extension theorems for Sobolev functions in Carnot-Carathéodory spaces*, J. Anal. Math., **74** (1998), 67-97.

[Gris] P. Grisvard, *Elliptic problems in non-smooth domains*, Pitman, Boston, (1985).

[J] D. Jerison, *The Dirichlet problem for the Kohn Laplacian on the Heisenberg group, Parts I and II*, J. Funct. Analysis, **43** (1981), no. 1, 97-142, no. 2, 224-257.

[JK] D. Jerison & C. K. Kenig, *The inhomogeneous Dirichlet problem in Lipschitz domains*, J. Funct. Anal., **130** (1995), 1, 161-219.

[JW] A. Jonsson & H. Wallin, *Function spaces on Subsets of* $\mathbb{R}^n$, Harwood Academic, Reading, UK, (1984).

[K] A. Kaplan, *Fundamental solutions for a class of hypoelliptic PDE generated by composition of quadratic forms*, Trans. Amer. Math. Soc., **258** (1980), 147-153.

[LiMa] J. L. Lion & M. Magenes, *Non-homogeneous boundary value problems and applications*, Vol. 1-2, Springer-Verlag, (1972).

[Me] M. Mekias, *Restriction to hypersurfaces of non-isotropic Sobolev spaces*, M.I.T. Ph.D Thesis, (1993).

[MM] R. Monti & D. Morbidelli, *Some trace theorems for vector fields*, preprint (2000).

[NSW] A. Nagel, E. M. Stein & S. Wainger, *Balls and metrics defined by vector fields I: basic properties*, Acta Math. **155** (1985), 103-147.

[Ro] C. Romero, *Potential theory for the Kohn Laplacian on the Heisenberg group*, Univ. of Minnesota Ph. D. Thesis (1991).

[RS] L. P. Rothschild & E. M. Stein, *Hypoelliptic differential operators and nilpotent groups*, Acta Math. **137** (1976), 247–320.

DEPARTMENT OF MATHEMATICS, THE JOHNS HOPKINS UNIVERSITY, BALTIMORE, MD 21218
E-mail address, Donatella Danielli: `dxd@math.jhu.edu`

DEPARTMENT OF MATHEMATICS, THE JOHNS HOPKINS UNIVERSITY, BALTIMORE, MD 21218
E-mail address, Nicola Garofalo: `ngarofalo@math.jhu.edu`

DEPARTMENT OF MATHEMATICS, GEORGETOWN UNIVERSITY, WASHINGTON DC 20057-1233
E-mail address, Duy-Minh Nhieu: `nhieu@math.georgetown.edu`

Contemporary Mathematics
Volume **277**, 2001

The Solution of the Kato Problem

Steve Hofmann

1. History of the Kato problem, and recent progress

In this note, we discuss some recently obtained joint results with P. Auscher, M. Lacey, J. L. Lewis, A. McIntosh, and P. Tchamitchian, in which we treat the so called "Kato problem" (or "square root problem") for second order divergence form operators.

Let A be an $n \times n$ matrix of complex, L^∞ coefficients, defined on $\mathbb{R}^n$, with $\|A\|_\infty \leq \Lambda$, and satisfying the ellipticity (or "accretivity") condition

$$\lambda|\xi^2| \leq \mathrm{Re}\langle A\xi, \xi\rangle \leq \Lambda|\xi|^2, \tag{1.1}$$

for $\xi \in \mathbb{C}^n$ and for some λ, Λ such that $0 < \lambda \leq \Lambda < \infty$. Here $\langle , \rangle$ denotes the usual inner product in $\mathbb{C}^n$, so that

$$\langle A\xi, \xi\rangle \equiv \sum_{i,j} A_{ij}(x)\xi_j \cdot \bar{\xi}_i.$$

We define a divergence form operator

$$Lu \equiv -\operatorname{div}(A(x)\nabla u), \tag{1.2}$$

which we interpret in the usual weak sense via a sesquilinear form. The accretivity condition (1.1) enables one to define an accretive square root $\sqrt{L} \equiv L^{1/2}$ (see [**14**]), and a fundamental issue is to "solve the square root problem", i.e. to establish the estimate

$$\|\sqrt{L}f\|_{L^2(\mathbb{R}^n)} \leq C\|\nabla f\|_{L^2(\mathbb{R}^n)}, \tag{1.3}$$

with C depending only on n, λ and Λ. The latter estimate is connected with the question of the analyticity of the mapping $A \to L^{\frac{1}{2}}$, which in turn has applications to the perturbation theory for certain classes of hyperbolic equations (see [**18**]). We note that it is well known, and easy to see, that (1.3) holds when L is self-adjoint. A long-standing open problem, essentially posed by Kato [**14**], (but refined by McIntosh [**18, 20**] - we shall explain this point more fully below) is the following:

I thank my collaborators P. Auscher, M. Lacey, J. L. Lewis, A. McIntosh, and P. Tchamitchian for their numerous and indispensable contributions to our joint work. Some of the historical material described in this article is taken from the introduction to my jointly authored paper [**2**], which will appear in Acta Mathematica. I am grateful to Acta Math. and to the Mittag-Leffler Institute for their kind permission to quote from that article. The author was supported by NSF.

QUESTION 1. *Let $A_z, z \in \mathbb{C}$ denote a family of accretive matrices as above, which in addition are holomorphic in z, and self-adjoint for real z. Let*

$$L_z \equiv -\operatorname{div} A_z(x)\nabla.$$

Is $L_z^{1/2}$ holomorphic in z, in a neighborhood of $z = 0$?

In fact, Kato actually formulated this question for a more general class of abstract accretive operators. A counterexample to the abstract problem was found by McIntosh [**20**]. However, it has been pointed out in [**18**] that, in posing the problem, Kato had been motivated by the special case of elliptic differential operators, and by the applicability of a positive result, in that special case, to the perturbation theory for hyperbolic evolution equations. A positive answer to the question posed above can be re-stated as follows:

CONJECTURE 1.4. *The estimate* (1.3) *holds in a complex neighborhood in L^∞ of any self-adjoint matrix A, satisfying* (1.1)*; i.e.* (1.3) *holds for the operator $\widetilde{L}$ (as in* (1.2)*) associated to any complex valued matrix $\widetilde{A}$, whenever*

$$\|\widetilde{A} - A\|_\infty \leq \epsilon_0,$$

with ϵ_0 depending only on n, λ and Λ.

Indeed, given Conjecture 1.4, then by the operator valued version of Cauchy's Theorem, one obtains analyticity at $z = 0$ of the mapping

$$z \longrightarrow L_z^{\frac{1}{2}},$$

where $L_z \equiv -\operatorname{div}(A_z)\nabla$, $z \to A_z$ is analytic, and $A_0 \equiv A$ is self-adjoint. It was this analyticity result that Kato had sought, in particular for real, symmetric matrices, in connection with the theory of hyperbolic equations. In [**14**], Kato also framed a more general conjecture for square roots of abstract accretive operators belonging to some broad class (see [**21**] for the details). Again the abstract question was shown to have a negative answer: a counter-example was obtained by McIntosh [**19**], who then reformulated the conjecture for the special case of elliptic differential operators:

CONJECTURE 1.5. *The estimate* (1.3) *holds for any operator L defined as in* (1.2)*, associated to an L^∞, $n \times n$ matrix A with complex entries, for which* (1.1) *holds.*

To establish the validity of Conjecture 1.5 has become known as the "Kato Problem", or "Square Root Problem". Until recently, both Conjecture 1.4 and Conjecture 1.5 had been proved completely only when $n = 1$. In the 1-dimensional case, the square root problem is essentially equivalent to the problem of establishing the L^2 boundedness of the Cauchy integral operator along a Lipschitz curve. Thus Conjecture 1.5, and hence also Conjecture 1.4, were proved in one dimension in the celebrated paper of Coifman, McIntosh and Meyer [**7**]. The precise nature of the relationship between the Cauchy integral operator along a Lipschitz curve, and the one dimensional Kato problem, was obtained in [**16**]. In higher dimensions, both Conjecture 1.4 and Conjecture 1.5 had been proved only in the case that A is close, in some sense, to a constant matrix (or in the case that one imposes some additional structure on the matrix—see [**5**] for some examples). The first result involving perturbations of constant matrices was due independently to Coifman, Deng and Meyer [**6**], and Fabes, Jerison and Kenig [**9**], who established the square root estimate (1.3) whenever $\|A - I\|_\infty \leq \epsilon(n)$. Clearly, their methods allowed one also to replace the identity matrix I by any constant accretive matrix, and this was certainly understood at that time (see [**10**]). Sharper bounds for the constant $\epsilon(n)$ on the order of $n^{-\frac{1}{2}}$ were obtained by Journé [**13**]. Another result in the same spirit was due to Fabes, Jerison, and Kenig [unpublished], who proved that an appropriate analogue of (1.3) holds when A is continuous (and hence, at least locally, close to a constant matrix). Extensions of these "small constant" results, with L^∞ replaced by BMO, and C replaced by VMO, were obtained by Escauriaza (VMO, unpublished), and by Auscher and Tchamitchian [**5**] (BMO with small norm; $ABMO$, a space somewhat beyond VMO, and, more generally, small perturbations of $ABMO$ in BMO). In the latter results, one still supposes that $A \in L^\infty$; the point is that the smallness of the perturbation is measured in a more general sense.

Recently, there has been significant progress on these problems. In the past year, first Conjecture 1.4, at least in the case that A is real, symmetric, and then very recently, and more generally, Conjecture 1.5, have been proved, along with extensions to systems of equations and to higher order operators. We state these results here; details will appear elswhere. The first theorem is joint work of P. Auscher, J.L. Lewis, P. Tchamitchian, and the author [**2**].

THEOREM 1.6. *Let $n \geq 1$. Suppose that A is a real, symmetric $n \times n$ matrix of L^∞ coefficients satisfying* (1.1). *Then there exists $\epsilon_0 \equiv \epsilon_0(n, \lambda, \Lambda)$ such that for any complex valued $n \times n$ matrix $\widetilde{A}$, with $\|A - \widetilde{A}\|_\infty \leq \epsilon_0$, the operator*

$$\widetilde{L} \equiv -\operatorname{div}(\widetilde{A}(x)\nabla)$$

satisfies (1.3), *with a constant C which depends only on n, λ, Λ. Moreover,*

$$\|\sqrt{\widetilde{L}}f - \sqrt{L}f\|_{L^2(\mathbb{R}^n)} \leq C(n, \lambda, \Lambda)\|A - \widetilde{A}\|_\infty \|\nabla f\|_{L^2(\mathbb{R}^n)}. \tag{1.7}$$

Thus we solve Conjecture 1.4 in all dimensions, at least in the case of perturbations of real symmetric matrices. The method of proof is based on a "Tb" theorem for square roots in [**5**], combined with an "extrapolation method" for Carleson measures, which utilizes a stopping time decomposition of a Carleson region in the spirit of Carleson's "corona" construction, and which was first developed to prove parabolic measure estimates [**17**], [**12**].

The next result, which extends the previous theorem, is joint work of M. Lacey, A. McIntosh and the author [**11**], in which we solve Conjecture 1.5 in all dimensions, assuming that the heat kernel $W_{t^2}(x, y)$, which is the kernel of the operator $e^{-t^2 L}$, satisfies the "Gaussian" property, i.e., assuming that there exist positive constants α and β such that

$$\begin{aligned} (1.8)\qquad & (i) \quad |W_{t^2}(x,y)| \leq \beta t^{-n} \exp\left\{\frac{-|x-y|^2}{\beta t^2}\right\} \\ & (ii) \quad |W_{t^2}(x+h,y) - W_{t^2}(x,y)| + |W_{t^2}(x,y+h) - W_{t^2}(x,y)| \\ & \qquad \leq \beta \frac{|h|^\alpha}{t^{n+\alpha}} \exp\left\{\frac{-|x-y|^2}{\beta t^2}\right\}, \end{aligned}$$

where the latter inequality holds whenever either $|h| \leq t$, or $|h| \leq |x - y|/2$. In particular, (1.8) always holds in two dimensions [**4**], and for perturbations of real elliptic matrices [**1**], in each case with α and β depending only on n, λ and Λ. We have the following ([**11**]):

THEOREM 1.9. *Suppose that A is an $n \times n$ matrix of L^∞, complex coefficients, defined on $\mathbb{R}^n$, such that* (1.1) *holds, and let $L \equiv -\operatorname{div}(A(x)\nabla)$. We suppose also that L satisfies property (G). Then the square root estimate* (1.3) *holds, with $C \equiv C(\alpha, \beta, n, \lambda, \Lambda)$.*

But the latter result is still not the final story. Indeed, P. Auscher, M. Lacey, A. McIntosh, P Tchamitchian and the present author will present the proof of Conjecture 1.5, in general ([**3**]). The point here is to remove the Gaussian hypothesis on the heat kernel. Our technique for doing so is based on the somewhat surprising fact, proved in [**3**], that a substitute for Gaussian bounds always holds, for any accretive second order divergence form elliptic operator: namely, that one can obtain exponential decay estimates for the resolvent kernel (or heat kernel) in the sense of L^2 averages. As a consequence we may then extend the arguments of [**11**], of course with some modifications owing to the lack of pointwise kernel bounds, to obtain the following:

THEOREM 1.10. *Let A be any $n \times n$ matrix of L^∞, complex coefficients, defined on $\mathbb{R}^n$, such that* (1.1) *holds, and let $L \equiv -\operatorname{div}(A(x)\nabla)$. Then the square root estimate* (1.3) *holds, with $C \equiv C(n, \lambda, \Lambda)$.*

The proofs of the latter two theorems are also based on the "Tb" criterion of [**5**], but using a different sort of stopping time approach in the application of the Tb theorem.

Finally, we remark that we have also obtained several extensions of these results. An analogue of Theorem 1.6 holds for square roots of systems, and Theorem 1.10 may be extended to the case of higher order elliptic operators, as well as to second order operators on Lipschitz domains. The details will appear in forthcoming papers.

References

[1] P. Auscher *Regularity theorems and heat kernels for elliptic operators*, J. London Math. Soc. **54** (1996), pp. 284-296.

[2] P. Auscher, S. Hofmann, J.L. Lewis, P. Tchamitchian, *Extrapolation of Carleson measures and the analyticity of Kato's square root operators*, to appear, Acta Math.

[3] P. Auscher, S. Hofmann, M. Lacey, A. McIntosh, and P. Tchamitchian, *The solution of the Kato square root problem for second order elliptic operators in $\mathbb{R}^n$*, preprint.

[4] P. Auscher, A. McIntosh and P. Tchamitchian, *Heat kernel of complex elliptic operators and applications,* J. Functional Analysis, **152** (1998), pp. 22-73.

[5] P. Auscher and P. Tchamitchian, *Square root problem for divergence operators and related topics*, Astérisque Vol. 249 (1998), Société Mathématique de France.

[6] R. Coifman, D. Deng, and Y. Meyer, *Domains de la racine carrée de certains operateurs differentials accrétifs*, Annales de l'Institut Fourier **33** (1983), 123-134.

[7] R. Coifman, A. McIntosh and Y. Meyer, *L'intégrale de Cauchy définit un opérateur borné sur L^2 pour les courbes lipschitziennes*, Annals of Math. **116** (1982), 361-387.

[8] R. Coifman and Y. Meyer, "Non-linear harmonic analysis, operator theory, and PDE," in *Beijing Lectures in Harmonic Analysis,* Annals of Math. Studies **112**, Princeton Univ. Press, Princeton, NJ, 1986, pp. 3-45.

[9] E. Fabes, D. Jerison and C. Kenig, *Multilinear square functions and partial differential equations*, Amer. J. of Math. **107** (1985), 1325-1367.

[10] E. Fabes, D. Jerison and C. Kenig, *Necessary and sufficient conditions for absolute continuity of elliptic-harmonic measure*, Annals of Math. **119** (1984), 121-141.

[11] S. Hofmann, M. Lacey and A. McIntosh, *The solution of the Kato problem for divergence form elliptic operators with Gaussian heat kernel bounds*, preprint.

[12] S. Hofmann and J.L. Lewis, *The Dirichlet problem for parabolic operators with singular drift terms*, to appear Memoirs of the Amer. Math. Soc.

[13] J.-L. Journé, *Remarks on the square root problem*, Pub. Math. **35** (1991), 299-321.
[14] T. Kato, *Fractional powers of dissipative operators*, J. Math. Soc. Japan **13** (1961), p. 246-274.
[15] T. Kato, *Integration of the equation of evolution in a Banach space*, J. Math. Soc. Japan **5** (1953), 208-234.
[16] C. Kenig and Y. Meyer, *The Cauchy integral on Lipschitz curves and the square root of second order accretive operators are the same*, in Recent Progress in Fourier Analysis, I. Peral, ed., North Holland Math Studies **111** (1985), 123-145.
[17] J.L. Lewis and M. Murray, *The method of layer potentials for the heat equation in time-varying domains*, Memoirs of the Amer. Math. Soc. **545** (1995).
[18] A.McIntosh, *Square roots of operators and applications to hyperbolic PDE*, in the proceedings of the Miniconference on Operator Theory and PDE, CMA, The Australian National University, Canberra, 1983.
[19] A. McIntosh, *On the Comparability of $A^{1/2}$ and $A^{*1/2}$*, Proc. Amer. Math. Soc. 32 (1972), 430-434.
[20] A. McIntosh, *On representing closed accretive sesquilinear forms as* $(A^{1/2}u, A^{*1/2}v)$, Collège de France Seminar, Volume III (H. Brezis and J.-L. Lions, eds.), Research Notes in Mathematics, no. 70, Pitman, 1982, pp. 252-267.
[21] A. McIntosh, *The square root problem for elliptic operators*, Functional Analytic Methods for Partial Differential Equations, Lecture Notes in Math. vol. 1450, Springer Verlag, 1990, pp. 122-140.

DEPARTMENT OF MATHEMATICS, UNIVERSITY OF MISSOURI-COLUMBIA, COLUMBIA, MO 65211
E-mail address: `hofmann@math.missouri.edu`

Contemporary Mathematics
Volume **277**, 2001

A DECOMPOSITION OF FUNCTIONS WITH VANISHING MEAN OSCILLATION

MICHAEL BRIAN KOREY

ABSTRACT. A function has vanishing mean oscillation (VMO) on $\mathbb{R}^n$ if its mean oscillation—the local average of its pointwise deviation from its mean value—both is uniformly bounded over all cubes within $\mathbb{R}^n$ and converges to zero with the volume of the cube. The more restrictive class of functions with vanishing lower oscillation (VLO) arises when the mean value is replaced by the minimum value in this definition. It is shown here that each VMO function is the difference of two functions in VLO.

1. INTRODUCTION

To say that a locally integrable function f on $\mathbb{R}^n$ has *bounded mean oscillation*, $f \in \text{BMO}$, means that

$$\sup_Q \frac{1}{|Q|} \int |f(x) - f_Q| \, dx < \infty, \tag{1}$$

where the supremum runs over all cubes Q in $\mathbb{R}^n$ with edges parallel to the coordinate axes, $|Q|$ denotes the measure of Q, and f_Q is the mean value of f over Q, i.e., $f_Q = (1/|Q|) \int_Q f$. A function f is said to have *bounded lower oscillation* if the term f_Q in (1) can be replaced by $\inf_Q f$, the essential infimum[1] of f over Q; in other words, $f \in \text{BLO}$ if

$$\sup_Q \frac{1}{|Q|} \int (f(x) - \inf f_Q) \, dx = \sup_Q (f_Q - \inf f_Q) < \infty. \tag{2}$$

The suprema in (1) and (2) are denoted by $\|f\|_{\text{BMO}}$ and $\|f\|_{\text{BLO}}$, respectively, and when only the cubes Q within a given cube Q_0 are considered, the symbols $\text{BMO}(Q_0)$ and $\text{BLO}(Q_0)$ will be used.

It is not difficult to see that each BLO function is in BMO; in fact, the estimate $\|f\|_{\text{BMO}} \leq 2\|f\|_{\text{BLO}}$ holds. Unlike BMO, however, the set BLO is not stable under multiplication by negative numbers ($\log|x|$ is in BLO, but $-\log|x|$ is not), and $\text{BLO} \cap (-\text{BLO}) = L^\infty$, as follows readily from the definition. On the other hand, Coifman and Rochberg [3], invoking a rather subtle argument of Carleson, showed that each BMO function can be written as the difference of two BLO functions, in short:

$$\text{BMO} = \text{BLO} - \text{BLO}.$$

1991 *Mathematics Subject Classification.* Primary 42B25; Secondary 26A15, 46E30.

Key words and phrases. Bounded mean oscillation, bounded lower oscillation, vanishing mean oscillation, A_p condition, uniform continuity.

[1]In general, all pointwise statements in this paper should be understood to hold only almost everywhere with respect to Lebesgue measure. In particular, $\inf_Q f$ and $\sup_Q f$ denote the essential infimum and essential supremum of f on Q, and no distinction will be made between two functions agreeing on the complement of a set of measure zero.

This is likewise a consequence of the factorization of A_p weights obtained by Jones [9];[2] in this regard, see also [2], [4], and [5].

The authors of [3] concluded their paper by noting that "many results about functions in BMO have analogs for functions of vanishing mean oscillation ... [but we] do not know what the analogs of our results are in that context." The purpose of this paper is to provide such an analog in the form of the following result:

Theorem. *Each* VMO *function is the difference of two* VLO *functions. That is, if* $f \in \mathrm{VMO}$, *then there exist* VLO *functions* F *and* G *such that* $f = F - G$ *and*

$$\|F\|_{\mathrm{BLO}} + \|G\|_{\mathrm{BLO}} \leq C\|f\|_{\mathrm{BMO}}.$$

The constant C *depends only on the dimension* n.

A word of explanation about the terms in the theorem is necessary. To say that a BMO function has *vanishing mean oscillation* means not only that the supremum in (1) is bounded over all cubes, but also that it vanishes in the asymptotic limit of ever smaller scales. In other words, $f \in \mathrm{VMO}$ if both $f \in \mathrm{BMO}$ and

$$\frac{1}{|Q|}\int |f(x) - f_Q|\,dx = o(1) \qquad (\ell(Q) \to 0),$$

where $\ell(Q)$ denotes the edge-length (or "size") of the cube Q. Each bounded, uniformly continuous (BUC) function is in VMO, uniform continuity being expressable in the form

$$\sup_Q f - \inf_Q f = o(1) \qquad (\ell(Q) \to 0), \tag{3}$$

and VMO is actaully the closure of BUC with respect to the BMO norm defined in (1), as was shown in [13]. The space VMO was introduced by by Sarason in [13] in the context of algebras on the unit circle and has since found manifold applications in diverse areas of analysis: [1] and [10] give two recent examples in the context of partial differential equations.

The new class VLO considered here, the set of functions with *vanishing lower oscillation*, consists of those BMO functions f for which

$$f_Q - \inf f_Q = o(1) \qquad (\ell(Q) \to 0). \tag{4}$$

VLO is a proper subset of VMO: while both $\sqrt{\log(1/|x|)}$ and $-\sqrt{\log(1/|x|)}$ are in VMO, only the former is in VLO. Moreover, the portion of VLO that is stable under multiplication by a negative number is a familiar space:

$$\mathrm{VLO} \cap (-\mathrm{VLO}) = \mathrm{BUC},$$

for when both f and $-f$ satisfy (4), then f also satisfies (3). What the theorem then says is that

$$\mathrm{VMO} = \mathrm{VLO} - \mathrm{VLO}.$$

Although the statement of the theorem is a straightforward generalization of that in [3], the proof is not. One essential reason for the difficulty is that while L^∞ functions are in BLO (and thus in BMO), they are not generally in VMO; the characteristic function of the unit interval $[0, 1]$ on the line, for example, has mean oscillation $1/2$ over each interval of the form $[-r, r]$, no matter how small r is. This presents a non-trivial obstacle to proving the theorem, because the techniques used

[2]In fact, $\mathrm{BMO} = \{\alpha \log w : \alpha \geq 0, w \in A_2\}$ and $\mathrm{BLO} = \{\alpha \log w : \alpha \geq 0, w \in A_1\}$, so that the statement $\mathrm{BMO} = \mathrm{BLO} - \mathrm{BLO}$ is the logarithm of the factorization result $A_2 = A_1/A_1$ in [9].

to characterize BMO in [3] and the Calderon-Zygmund decomposition as applied in [9] typically lead to bounded remainder terms. In the decomposition of BMO, this level of precision suffices and such remainder terms need not be broken down any further; on the other hand, if VMO functions are to be decomposed, then attention must be paid not only to the size but also to the smoothness of these remainders. (A similar difficulty is encountered when trying to extend VMO functions defined on a measurable set to all of $\mathbb{R}^n$, as in [7].)

The method of proof for the VMO theorem stated above is first to show the result in the dyadic model case, applying ideas used by Jones [9] for the factorization of Muckenhoupt A_p weights. Translations of this dyadic result will then be averaged to give the general case, along the lines of of Garnett and Jones [6]. The same combination of techniques was also used by the present author in [11]. There the focus was on showing that each Muckenhoupt weight with small A_p bound—that is, with bound near 1, the bound being *fixed and valid over all scales*—could be factored as a product of suitable powers of two A_1 weights with weight bounds also close to 1.[3] Here, by contrast, the challenge is to decompose a given VMO function—a BMO function whose oscillatory behavior *improves over ever smaller scales* in an asymptotically optimal manner—as the difference of two simpler functions that also display this asymptotically optimal improvement.

2. The dyadic setting

The first goal is to prove a dyadic form of the decomposition theorem. This model version is stated for the collection $\mathcal{D}(Q_0)$ of all dyadic subcubes of an arbitrary, fixed cube Q_0 in $\mathbb{R}^n$, that is, all those cubes obtained by dividing Q_0 into 2^n congruent cubes of half its length, dividing each of these into 2^n congruent cubes, and so on; by convention, Q_0 itself belongs to $\mathcal{D}(Q_0)$. The space $\mathrm{BMO}^{\mathrm{dy}}(Q_0)$ then denotes the set of all integrable functions on Q_0 for which condition (1) holds when the supremum there is taken only over cubes Q in $\mathcal{D}(Q_0)$; dyadic versions of BLO, VMO, VLO, and BUC are defined analogously.

At first, we derive a preliminary result (Lemma 1) valid only on large dyadic scales; with the help of a suitably localized iteration procedure, we shall then extend this to the full dyadic version (Lemma 2).

Lemma 1. *Suppose* $f \in \mathrm{BMO}^{\mathrm{dy}}(Q_0)$. *Let* δ *be a dyadic scale (i.e.,* $\delta \in 2^{-\mathbb{N}}$*). Then there exist bounded functions* F_1, G_1 *and* R_1, *all locally constant*[4] *on the mesh* $\{Q_1 \in \mathcal{D}(Q_0) : \ell(Q_1) = \delta\ell(Q_0)\}$ *of dyadic subcubes of relative size* δ, *satisfying both the pointwise identity*

$$f(x) - f_{Q_0} = F_1(x) - G_1(x) + R_1(x) + \sum_{\ell(Q_1)=\delta\ell(Q_0)} (f(x) - f_{Q_1})\chi_{Q_1}(x) \qquad (x \in Q_0)$$

and the estimate

$$\|F_1\|_{\mathrm{BLO}^{\mathrm{dy}}(Q_0)} + \|G_1\|_{\mathrm{BLO}^{\mathrm{dy}}(Q_0)} + \|R_1\|_{L^\infty(Q_0)} \le C\|f\|_{\mathrm{BMO}(Q_0)}. \tag{5}$$

The constant C *depends only on the dimension* n.

---

[3]The sharp result is that a weight with A_2 bound $1+\varepsilon$ can be factored as the quotient of two A_1 weights with bounds $1 + \mathcal{O}(\sqrt{\varepsilon})$, as $\varepsilon \to 0$; see [11] for details.

[4]Local constancy on a mesh of congruent cubes simply means constancy on each cube within the mesh.

The proof uses an iterative Calderón-Zygmund decomposition to single out those cubes on which the mean oscillation of f is large. Note that in this first lemma only cubes of comparable size enter—here all dyadic Q with edges of length between $\delta\ell(Q_0)$ and $\ell(Q_0)$; in Lemma 2, the full range of dyadic scales will be considered, and special attention will be paid to how the oscillatory behavior of VMO functions improves as the scale decreases.

Proof. Without loss of generality, let $Q_0 = [0,1]^n$ and $\|f\|_{\mathrm{BMO}^{\mathrm{dy}}(Q_0)} = 2^{-n}$. As is well known, the BMO condition strongly restricts how the average values of a function vary with the size of the averaging set. In particular, for any cube Q and any other cube $2Q$ containing it and with edges twice as long,

$$|f_Q - f_{2Q}| \leq 2^n \frac{1}{|2Q|} \int_{2Q} |f - f_{2Q}|. \tag{6}$$

When $f \in \mathrm{BMO}$, the right-hand side is bounded; when $f \in \mathrm{VMO}$, the right-hand side vanishes as $\ell(Q) \to 0$. This simple observation drives the stopping-time argument below.

Let $\mathcal{G}_1^0 = \{Q_0\}$. Inequality (6) guarantees that $|f_Q - f_{Q_0}| \leq 1$, when Q is any one of the 2^n subcubes of Q_0 obtained from it by bisecting its edges. Fix $\delta \in 2^{-\mathbb{N}}$. Of interest are the largest subcubes Q obtained by further bisection on which $|f_Q - f_{Q_0}| > 1$ (and $\ell(Q) > \delta$). In detail, define

$$\mathcal{G}_1^1 = \{Q \in \mathcal{D}(Q_0) : |f_Q - f_{Q_0}| > 1,\ \ell(Q) > \delta,\ Q\,\mathrm{maximal}\} \tag{7}$$

and, inductively,

$$\mathcal{G}_1^{m+1} = \{Q \in \mathcal{D}(Q') : Q' \in \mathcal{G}_1^m,\ |f_Q - f_{Q'}| > 1,\ \ell(Q) > \delta,\ Q\,\mathrm{maximal}\}. \tag{8}$$

Write $\mathcal{G}_1 = \bigcup_{m=0}^{\infty} \mathcal{G}_1^m$ for the full collection of all such selected cubes. Let Ω_1^m be the union of the cubes in $\mathcal{G}_1^m$; then, by construction, $\Omega_1^{m+1} \subseteq \Omega_1^m \subseteq \cdots \subseteq \Omega_1^0$. For each proper dyadic subcube Q of $\mathcal{D}(Q_0)$ with $\ell(Q) \geq \delta$, let $\widetilde{Q}$ denote the minimal cube in $\mathcal{G}_1$ that strictly contains it;[5] this means, in the special case when $Q \in \mathcal{G}_1^{m+1}$, that $\widetilde{Q}$ is the unique cube in $\mathcal{G}_1^m$ containing Q.

Each dyadic subcube Q_1 of edge-length δ is then contained in a unique decreasing chain of selected cubes of the form[6]

$$Q_0 \supseteq Q_1^1 \supseteq Q_1^2 \supseteq \cdots \supseteq Q_1^m = \widetilde{Q}_1 \qquad (Q_1^i \in \mathcal{G}_1^i,\ 1 \leq i \leq m).$$

The difference $f(x) - f_{Q_0}$ can be expanded as a telescoping sum over pairs of adjacent cubes in this chain, extended by the additional inclusion $\widetilde{Q}_1 \supset Q_1$. What results from this is the pointwise identity

$$\begin{aligned} f(x) - f_{Q_0} = \sum_{m=1}^{\infty} \sum_{Q \in \mathcal{G}_1^m} (f_Q - f_{\widetilde{Q}})\chi_Q(x) \\ + (f_{Q_1} - f_{\widetilde{Q}_1}) + f(x) - f_{Q_1} \qquad (x \in Q_1), \end{aligned} \tag{9}$$

which holds on each dyadic $Q_1 \in \mathcal{D}(Q_0)$ of size δ.

---

[5]That is, $\widetilde{Q} = \bigcap\{Q' \in \mathcal{G}_1 : Q \subset Q'\}$.

[6]The length of the chain is at least 1 (in the extreme case when no predecessor of Q_1 is selected, so that $\widetilde{Q}_1 = Q_0$) and at most $1 + \log_2(1/\delta)$; the actual length varies over the mesh of dyadic subcubes Q_1 within Q_0 of size δ. Accordingly, all but finitely many of the sets $\mathcal{G}_1^m$ are empty, so that the double sum in (9) is actually a sum over only finitely many cubes.

How large are the terms in (9)? Maximality in the selection criteria (7) and (8) and the basic BMO estimate (6) give rise to the mean-value inequality

$$1 < |f_Q - f_{\widetilde{Q}}| \le 2 \qquad \Big(Q \in \bigcup_{m=1}^{\infty} \mathcal{G}_1^m\Big). \tag{10}$$

They also lead to the relative density estimate

$$|Q \cap \Omega_1^{m+1}| \le 2^{-n}|Q| \qquad (Q \in \mathcal{G}_1^m), \tag{11}$$

which is valid for each $m \ge 0$. Summing this last estimate over all the cubes Q in $\mathcal{G}_1^m$ and iterating leads to the bound

$$|\Omega_1^m| \le 2^{-mn}|Q_0| \qquad (m \in \mathbb{N}). \tag{12}$$

Next, to obtain suitable dyadic BLO summands of f, split the double sum in (9) according to the sign of the difference $f_Q - f_{\widetilde{Q}}$. Then

$$f(x) - f_{Q_0} = F_1(x) - G_1(x) + R_1(x) + \sum_{\ell(Q_1)=\delta} (f(x) - f_{Q_1})\chi_{Q_1}(x) \qquad (x \in Q_0),$$

where

$$F_1(x) = \sum_{m=1}^{\infty} \sum_{Q \in \mathcal{G}_1^m} (f_Q - f_{\widetilde{Q}})^+ \chi_Q(x), \tag{13}$$

$$G_1(x) = \sum_{m=1}^{\infty} \sum_{Q \in \mathcal{G}_1^m} (f_{\widetilde{Q}} - f_Q)^+ \chi_Q(x), \tag{14}$$

and

$$R_1(x) = \sum_{\ell(Q_1)=\delta} (f_{Q_1} - f_{\widetilde{Q}_1})\chi_{Q_1}(x).$$

It is important to note that the functions F_1 and G_1 defined in (13) and (14) are non-negative; where they are positive, their value must, by (10), exceed 1.

Proving the L^∞ estimate in (5) for R_1 is straightforward: since, by the definition of $\widetilde{Q}_1$, no dyadic cube between Q_1 and $\widetilde{Q}_1$ is selected in the stopping-time argument (in other words, $|f_Q - f_{\widetilde{Q}_1}| \le 1$ whenever $Q_1 \subset Q \subset \widetilde{Q}_1$), then either $|f_{Q_1} - f_{\widetilde{Q}_1}| \le 1$ or $1 < |f_{Q_1} - f_{\widetilde{Q}_1}| \le 2$. In either case, $|R_1(x)| \le 2$ for a.e. $x \in Q_0$, as desired.

As F_1 and G_1 are locally constant on the mesh $\{Q_1 \in \mathcal{D}(Q_0) : \ell(Q_1) = \delta\}$, it remains to estimate their (dyadic) BLO bounds on cubes larger than those in this mesh. To confirm (5), then, we must show, for each $Q \in \mathcal{D}(Q_0)$ with $\ell(Q) > \delta$, that there exists a constant C_Q (with $\sup_Q C_Q < \infty$) such that both

$$(F_1)_Q \le C_Q + \inf_Q F_1 \tag{15}$$

and

$$(G_1)_Q \le C_Q + \inf_Q G_1. \tag{16}$$

To prove this we now consider three cases.

Case I: The initial cube. We first verify (15) in the case when $Q = Q_0$, the original cube. In this case, $\inf_Q F_1 = 0$, for the choice of the height 1 in the stopping-time argument defining the selection procedure (7) ensures that the set $Q_0 \setminus \Omega_1^1$ has positive measure; see (12). Since $\int_Q F_1 = \int_0^\infty |E_t|\,dt$ with $E_t = \{x \in Q : F_1(x) > t\}$, estimating the dyadic VLO bound of F_1 then reduces to estimating the measure of the set E_t. But condition (10) ensures that $E_t \subseteq \Omega_1^1$, when $0 \le t < 2$, and, in general, that $E_t \subseteq \Omega_1^k$, when $k \in \mathbb{N}$ and $2(k-1) \le t < 2k$. Thus, by (12),

$$(F_1)_Q = \sum_{k=1}^{\infty} \int_{2(k-1)}^{2k} \frac{|E_t|}{|Q|}\,dt \le 2\sum_{k=1}^{\infty} \frac{|\Omega_1^k|}{|Q|} \le 2\sum_{k=1}^{\infty} 2^{-nk} \le 2 + \inf_Q F_1, \tag{17}$$

which is (15) for $Q = Q_0$.

Case II: A cube with a large jump in mean value. Suppose now that $Q \in \mathcal{G}_1^m$ for some positive m and that $f_Q - f_{\widetilde{Q}} > 1$.[7] Then $(F_1)_Q - \inf_Q F_1 = |Q|^{-1} \int_0^\infty |\widetilde{E}_t|\,dt$, where $\widetilde{E}_t = \{x \in Q : F_1(x) - \inf_Q F_1 > t\}$. In analogy to the first case, we find from (10) and (11) that $\widetilde{E}_t \subset Q \cap \Omega_1^{m+k}$, when $k \in \mathbb{N}$ and $2(k-1) \le t < 2k$. So for t in this range, $|\widetilde{E}_t| \le 2^{-nk}|Q|$, from which the desired estimate (15) once again follows, with $C_Q = 2$.

Case III: Cubes with no large jump in the mean. In Case I, we considered Q_0; in Case II, we treated those dyadic cubes Q within Q_0 (of length exceeding δ) for which $f_Q - f_{\widetilde{Q}} > 1$. To handle the remaining case efficiently, a bit of further notation is helpful. Recall that, for each proper dyadic subcube Q of Q_0 with $\ell(Q) \ge \delta$, the symbol $\widetilde{Q}$ denotes the minimal cube in $\mathcal{G}_1 = \cup_{m=1}^\infty \mathcal{G}_1^m$ that strictly contains Q; now set

$$\mathcal{P}(Q) = \{I \in \mathcal{D}(Q) : f_I - f_{\widetilde{Q}} > \ \ 1,\ \ell(I) > \delta,\ I\,\text{maximal}\},$$
$$\mathcal{N}(Q) = \{I \in \mathcal{D}(Q) : f_I - f_{\widetilde{Q}} < -1,\ \ell(I) > \delta,\ I\,\text{maximal}\}.$$

Note that the union of $\mathcal{P}(Q)$ and $\mathcal{N}(Q)$ is exactly the set of the cubes in $\cup_{m=1}^\infty \mathcal{G}_1^m$ that lie within Q. In this notation, the remaining case now consists of proving (15) on each dyadic cube Q (of size exceeding δ) for which $Q \notin \mathcal{P}(Q)$.

Fix such a cube Q. To estimate F_1, split Q into the union of its subcubes in $\mathcal{P}(Q)$ and the complement of this union. On the one hand, if $I \in \mathcal{P}(Q)$, then $\widetilde{I} = \widetilde{Q}$; Case II then applies, so that $\int_I F_1 \le (2 + \inf_I F_1)|I|$. Moreover,

$$1 < \inf_I F_1 - \inf_{\widetilde{Q}} F_1 = f_I - f_{\widetilde{I}} \le 2,$$

by (10). On the other hand, on the complement in Q of $\cup_{I\in\mathcal{P}(Q)} I$ the value of F_1 is exactly $\inf_{\widetilde{Q}} F_1$. In sum, then,

$$\begin{aligned}\int_Q F_1 &\le \sum_{I\in\mathcal{P}(Q)} (2 + \inf_I F_1)|I| + (\inf_{\widetilde{Q}} F_1)|Q \setminus \cup_{I\in\mathcal{P}(Q)} I| \\ &\le (4 + \inf_{\widetilde{Q}} F_1) \sum_{I\in\mathcal{P}(Q)} |I| + (\inf_{\widetilde{Q}} F_1)|Q \setminus \cup_{I\in\mathcal{P}(Q)} I| \\ &\le (4 + \inf_{\widetilde{Q}} F)|Q|.\end{aligned}$$

[7]Unlike in (8), the sign of the difference is important here.

Since $\inf_{\tilde{Q}} F_1 \le \inf_Q F_1$, the bound (15) with $C_Q = 4$ thus also holds for the cubes Q in this, the last case.

The justification of the dyadic BLO bound (16) is similar, with G_1 in place of F_1, $\mathcal{N}(Q)$ in place of $\mathcal{P}(Q)$, etc. This completes the proof of Lemma 1. □

When a function has (dyadic) vanishing mean oscillation, then we can apply the previous lemma in an iterative way that takes advantage of the function's improved behavior at small scales. This is the essence of the next result.

Lemma 2. *Suppose* $f \in \mathrm{VMO}^{\mathrm{dy}}(Q_0)$. *Then there exist non-negative functions* $F, G \in \mathrm{VLO}^{\mathrm{dy}}(Q_0)$ *and a function* $R \in \mathrm{BUC}^{\mathrm{dy}}(Q_0)$ *such that*

$$f(x) - f_{Q_0} = F(x) - G(x) + R(x) \qquad (x \in Q_0). \tag{18}$$

Writing $f(x)$ as the difference $(F(x) + R(x) + f_{Q_0}) - G(x)$ thus gives the dyadic form of the decomposition theorem.

Proof. As above, let $\|f\|_{\mathrm{BMO}^{\mathrm{dy}}(Q_0)} = 2^{-n}$. Since $f \in \mathrm{VMO}^{\mathrm{dy}}(Q_0)$, we can find a strictly decreasing sequence of dyadic scales on which the mean oscillation of f vanishes exponentially fast; that is, set $\delta_0 = \ell(Q_0)$ and choose $\{\delta_j\}_{j\in\mathbb{N}} \subset 2^{-\mathbb{N}}$ such that both $\delta_{j+1} < \delta_j$ (hence $\delta_j \to 0$ as $j \to \infty$) and

$$\frac{1}{|Q|}\int_Q |f(x) - f_Q|\, dx \le 2^{-n-j} \qquad (Q \in \mathcal{D}(Q_0), \ell(Q) \le \delta_j). \tag{19}$$

Further, let $Q_j(x)$ denote the dyadic subcube of Q_0 that contains x and has edge-length δ_j; then $\{Q_j(x) : x \in Q_0\}$ is a family of non-overlapping, congruent subcubes whose union is Q_0, and $Q_j(x) \to \{x\}$ as $j \to \infty$ for a.e. x. (When no confusion can arise, we shall simply write Q_j for $Q_j(x)$.)

The previous lemma (with $\delta = \delta_1/\delta_0$) gives rise to the decomposition

$$f(x) - f_{Q_0} = F_1(x) - G_1(x) + R_1(x) + \sum_{\ell(Q_1)=\delta_1} (f(x) - f_{Q_1})\chi_{Q_1}(x) \qquad (x \in Q_0)$$

as well as to a suitable estimate on the terms F_1, R_1, and G_1. Now apply the lemma to f on each cube Q_1 of size δ_1 (with $\delta = \delta_2/\delta_1$) in place of Q_0. Then

$$f(x) - f_{Q_1} = F_2(x) - G_2(x) + R_2(x) + \sum_{\ell(Q_2)=\delta_2} (f(x) - f_{Q_2})\chi_{Q_2}(x) \qquad (x \in Q_1).$$

Repeat the procedure, applying the lemma on each cube Q_{j-1} of size δ_{j-1} (with $\delta = \delta_j/\delta_{j-1}$) to obtain[8]

$$f(x) - f_{Q_{j-1}} = F_j(x) - G_j(x) + R_j(x) + \sum_{\ell(Q_j)=\delta_j} (f(x) - f_{Q_j})\chi_{Q_j}(x) \qquad (x \in Q_{j-1}).$$

[8]For each scale index $j \ge 0$, let $\mathcal{G}_j^0 = \{Q_{j-1} \in \mathcal{D}(Q_0) : \ell(Q_{j-1}) = \delta_{j-1}\}$ denote the full mesh of subcubes of size δ_{j-1}. Then, for $m \ge 0$, the above procedure inductively defines $\mathcal{G}_j^{m+1}$ from $\mathcal{G}_j^m$ by the rule: $\mathcal{G}_j^{m+1} = \{Q \in \mathcal{D}(Q') : Q' \in \mathcal{G}_j^m,\ |f_Q - f_{Q'}| > 2^{1-j},\ \ell(Q) > \delta_j,\ Q\,\text{maximal}\}$. In addition, when $Q \in \mathcal{D}(Q_0)$ and $\delta_j < \ell(Q) \le \delta_{j-1}$, then $\tilde{Q}$ denotes the minimal cube in $\mathcal{G}_j = \cup_{m=0}^{\infty}\mathcal{G}_j^m$ that properly contains Q.

The functions F_j, G_j, and R_j so obtained[9] are all locally constant on the mesh of dyadic subcubes of Q_0 of size δ_j. On account of (5) and (19) they also satisfy the estimate

$$\|F_j\|_{\mathrm{BLO}^{\mathrm{dy}}(Q_{j-1})} + \|G_j\|_{\mathrm{BLO}^{\mathrm{dy}}(Q_{j-1})} + \|R_j\|_{L^\infty(Q_{j-1})} \leq C2^{-j}. \tag{20}$$

uniformly on all dyadic subcubes Q_{j-1} within Q_0 of size δ_{j-1}.

All together, the first J iterations of the procedure just described yield

$$f(x) - f_{Q_0} = \sum_{j=1}^{J} F_j(x) - \sum_{j=1}^{J} G_j(x) + \sum_{j=1}^{J} R_j(x) + \sum_{\ell(Q_J)=\delta_J} (f(x) - f_{Q_J})\chi_{Q_J}(x),$$

which is valid for $x \in Q_0$. Since $\lim_{J\to\infty} f_{Q_J(x)} \to f(x)$ for a.e. x, the last sum vanishes in the limit. As a result,

$$\begin{aligned} f(x) - f_{Q_0} &= \sum_{j=1}^{\infty} F_j(x) - \sum_{j=1}^{\infty} G_j(x) + \sum_{j=1}^{\infty} R_j(x) \\ &= F(x) - G(x) + R(x) \qquad (x \in Q_0), \end{aligned} \tag{21}$$

which is (18). What remains is to show the appropriate estimates for the functions F, G, and R, so defined.

For this, fix an arbitrary $Q \in \mathcal{D}(Q_0)$ and find the unique number $J \in \mathbb{N}$ such that $\delta_J < \ell(Q) \leq \delta_{J-1}$. On Q, the functions F_j, G_j, and R_j $(1 \leq j \leq J-1)$ are then all constant. Hence

$$F_Q - \inf_Q F = \left(\sum_{j=J}^{\infty} F_j\right)_Q - \inf_Q \sum_{j=J}^{\infty} F_j \leq \sum_{j=J}^{\infty} ((F_j)_Q - \inf_Q F_j) \tag{22}$$

and

$$\sup_Q R - \inf_Q R = \sup_Q \sum_{j=J}^{\infty} R_j - \inf_Q \sum_{j=J}^{\infty} R_j \leq \sum_{j=J}^{\infty} 2 \sup_Q |R_j|.$$

On account of (20), the latter estimate on the (dyadic) modulus of continuity of R becomes

$$\sup_Q R - \inf_Q R \leq 2C \sum_{j=J}^{\infty} (2^{-j}) \qquad (Q \in \mathcal{D}(Q_0), \ell(Q) \leq \delta_{J-1}). \tag{23}$$

Since $\sum_{j=J}^{\infty} 2^{-j} \to 0$ as $J \to \infty$ (i.e., as $\delta_J \to 0$), then $R \in \mathrm{BUC}^{\mathrm{dy}}(Q_0)$.

The corresponding estimate for F is a bit more involved. The key is to determine how large each difference $(F_j)_Q - \inf_Q F_j$ in (22) can be. Recall that $\delta_J < \ell(Q) \leq \delta_{J-1}$. When $j = J$, then $(F_j)_Q - \inf_Q F_j \leq C2^{-j}$, by (20), since Q is contained in some dyadic cube Q_{J-1} of size δ_{J-1}.[10] Next, when $j \geq J+1$, then Q is a finite union of non-overlapping dyadic cubes Q_{j-1} of size δ_{j-1}. On each of these, the selection procedure guarantees that $\inf_{Q_{j-1}} F_j = 0$ and that $(F_j)_{Q_{j-1}} \leq C2^{-j}$.[11] Over all of Q, then, $\inf_Q F_{j-1} = 0$ and (taking the average

[9]Strictly speaking, the argument yields functions F_j, G_j, and R_j that are defined separately on each dyadic subcube Q_{j-1} of size δ_{j-1}; however, as the cubes $\{Q_{j-1}(x) : x \in Q_0\}$ of this size are non-overlapping and together cover Q_0, we shall view these functions as being defined a.e. on all of Q_0. Hence, the statement $\|R_j\|_{L^\infty(Q_{j-1})} \leq C2^{-j}$ (for all such Q_{j-1}) is equivalent to the global bound $\|R_j\|_{L^\infty(Q_0)} \leq C2^{-j}$.

[10]In other words, one of the Cases I–III arising in the proof of Lemma 1 applies to F_J on Q.

[11]This corresponds to Case I for F_j on Q_{j-1}.

of the mean values $(F_j)_{Q_{j-1}}$ over the Q_{j-1} comprising Q) also $(F_j)_Q \leq C2^{-j}$. In sum,

$$(F_j)_Q - \inf_Q F_j \leq C2^{-j} \qquad (Q \in \mathcal{D}(Q_0)) \tag{24}$$

for each $j \in \mathbb{N}$. Combining (22) and (24) gives the BLO bound

$$F_Q - \inf_Q F \leq C \sum_{j=J}^{\infty} (2^{-j}) \qquad (Q \in \mathcal{D}(Q_0),\, \ell(Q) \leq \delta_{J-1}).$$

The estimate for the BLO bound of G runs the same way. Letting $J \to \infty$ then shows that $F, G \in \mathrm{VMO}^{\mathrm{dy}}(Q_0)$, as claimed. □

Several remarks on the construction are in order; for simplicity, these are stated under the assumption that $\|f\|_{\mathrm{BMO}^{\mathrm{dy}}(Q_0)} \leq 2^{-n}$. First, observe that the functions F and G (resp. R) can be expressed as sums over all the dyadic subcubes of Q_0, not just over those where the mean oscillation of f is large (resp. not just over all dyadic subcubes having size δ_j). This means that

$$F(x) = \sum_{j=1}^{\infty} \sum_{m=1}^{\infty} \sum_{Q \in \mathcal{G}_j^m} (f_Q - f_{\tilde{Q}})^+ \chi_Q(x) = \sum_{Q_k \in \mathcal{D}(Q_0)} a_k \chi_{Q_k}(x), \tag{25}$$

$$G(x) = \sum_{j=1}^{\infty} \sum_{m=1}^{\infty} \sum_{Q \in \mathcal{G}_j^m} (f_{\tilde{Q}} - f_Q)^+ \chi_Q(x) = \sum_{Q_k \in \mathcal{D}(Q_0)} b_k \chi_{Q_k}(x), \tag{26}$$

and

$$R(x) = \sum_{j=1}^{\infty} \sum_{\ell(Q_j)=\delta_j} (f_{Q_j} - f_{\tilde{Q}_j}) \chi_{Q_j}(x) = \sum_{Q_k \in \mathcal{D}(Q_0)} c_k \chi_{Q_k}(x), \tag{27}$$

where the latter sum in each line runs over all of $\mathcal{D}(Q_0)$. In (25), for example, whenever $Q_k \notin \cup_{j=1}^{\infty} \cup_{m=1}^{\infty} \mathcal{G}_j^m$ or whenever $Q_k \in \cup_{m=1}^{\infty} \mathcal{G}_j^m$ for some j but $f_{Q_k} - f_{\tilde{Q}_k} < 0$, then $a_k = 0$; otherwise, $a_k = f_{Q_k} - f_{\tilde{Q}_k} \in (2^{1-j}, 2^{2-j}]$. A similar statement applies to the coefficients b_k. In (27), $c_k = (f_{Q_k} - f_{\tilde{Q}_k}) \in [-2^{2-j}, 2^{2-j}]$ if $\ell(Q_k) = \delta_j$ (for some $j \in \mathbb{N}$), whereas $c_k = 0$ otherwise. This reformulation will prove useful below.

The second remark concerns the rate at which the mean oscillation of a given VMO function vanishes; in the notation above, this is captured by the relative size of the sequence $\{\delta_j\}$ of scales in (19). As was noted in [14], each function in VMO is actually in BMO_φ—meaning $(1/|Q|) \int_Q |f(x) - f_Q|\, dx < \varphi(|Q|)$ for all Q—for some bounded, continuous, non-decreasing function φ on $\mathbb{R}_+$ having $\varphi(0) = 0$. Building on the work of Campanato and Meyers, Spanne [15] showed that BMO_φ functions are essentially uniformly continuous just in case the Dini condition $\int_0^1 \varphi(t)/t\, dt < \infty$ is fulfilled; this corresponds to the condition $\sum_{j=1}^{\infty} 2^{-j} \log(\delta_{j-1}/\delta_j) < \infty$ in the notation used here. Shi and Torchinsky took this analysis one step further in [14] and gave an elegant proof of a version of the John-Nirenberg inequality for BMO_φ. Their iterative Calderón-Zygmund decomposition has much in common with that used in the proof of the dyadic version of the theorem above (Lemmas 1 and 2 taken together), with the exception that it disregards bounded remainder terms and focuses only on the size, not the sign, of the jumps arising in the stopping-time argument. Precisely these two points, fine control over the remainders and a partitioning of the jumps by sign, are key features of the argument given above:

the former explains why a Calderón-Zygmund decomposition has been applied to all cubes in each mesh $\{Q_j \in \mathcal{D}(Q_0) : \ell(Q_j) = \delta_j\}$, rather than just to those having predecessors ("dyadic ancestors") that were selected within some coarser mesh, and the latter provides the basis for splitting the VMO function f into the VLO summands F and G.

3. The general setting

The proof of the theorem follows the argument in [6, pp. 361–64], except for certain technical modifications which are introduced to reflect the improved oscillatory behavior of VMO functions over small scales. For completeness, the full proof is given here. Let S_N denote the cube $\left[-2^N, 2^N\right]^n$.

Lemma 3. *Suppose f is in* VMO. *Then, for each $N \in \mathbb{N}$, there exist non-negative functions $\bar{F}_N, \bar{G}_N \in \mathrm{VLO}(S_N)$ and a function $\bar{R}_N \in \mathrm{BUC}(S_N)$ satisfying both the identity*

$$f(x) - f_{S_N} = \bar{F}_N(x) - \bar{G}_N(x) + \bar{R}_N(x) \qquad (x \in S_N) \tag{28}$$

and the estimate

$$\|\bar{F}_N\|_{\mathrm{BLO}(S_N)} + \|\bar{G}_N\|_{\mathrm{BLO}(S_N)} + \|\bar{R}_N\|_{L^\infty(S_N)} \leq C\|f\|_{\mathrm{BMO}}. \tag{29}$$

The constant C depends only on the dimension n.

Moreover, for each $M \in \mathbb{N}$, the family $\{\bar{R}_N : N \geq M\}$ is equicontinuous on S_M and the families $\{\bar{F}_N : N \geq M\}$ and $\{\bar{G}_N : N \geq M\}$ have uniformly vanishing lower oscillation on S_M; this means that the quantity $C_Q = C_Q(M)$ defined by

$$C_Q = \sup_{N \geq M} \left((\bar{F}_N)_Q - \inf_Q \bar{F}_N, (\bar{G}_N)_Q - \inf_Q \bar{G}_N \right) \tag{30}$$

is finite for each $Q \subseteq S_M$ and that $C_Q \to 0$ as $\ell(Q) \to 0$.

Note that the estimates on the lower oscillation of $\bar{F}_N$ and $\bar{G}_N$ are asserted to hold over all subcubes of S_N, not just the dyadic ones; the uniform continuity of $\bar{R}_N$ on S_N is also meant in the standard sense (without respect to any dyadic mesh).

Let us first show how this last lemma implies the theorem. The identity (28) can be re-written, after subtracting off the mean value of each side on S_0, as

$$\begin{aligned} f(x) - f_{S_0} &= [\bar{F}_N(x) - (\bar{F}_N)_{S_0}] - [\bar{G}_N(x) - (\bar{G}_N)_{S_0}] + [\bar{R}_N(x) - (\bar{R}_N)_{S_0}] \\ &= \tilde{F}_N(x) - \tilde{G}_N(x) + \tilde{R}_N(x). \end{aligned} \tag{31}$$

Then (29), in conjunction with the John-Nirenberg inequality of [8], yields a uniform bound on the quadratic mean oscillation of $\tilde{F}_N$,[12] namely

$$\frac{1}{|Q|} \int_Q |\tilde{F}_N - (\tilde{F}_N)_Q|^2 \leq C\|f\|_{\mathrm{BMO}}^2 \qquad (Q \subseteq S_N).$$

Suppose now that $M \leq N$. When $Q = S_M$, the last line becomes

$$\frac{1}{|S_M|} \int_{S_M} |\tilde{F}_N|^2 \leq C\|f\|_{\mathrm{BMO}}^2 + |(\tilde{F}_N)_{S_M}|^2. \tag{32}$$

To control the right-hand side here, form a telescoping sum of mean values:

$$(\tilde{F}_N)_{S_M} = (\tilde{F}_N)_{S_0} + [(\tilde{F}_N)_{S_1} - (\tilde{F}_N)_{S_0}] + \cdots + [(\tilde{F}_N)_{S_M} - (\tilde{F}_N)_{S_{M-1}}]. \tag{33}$$

[12] The estimate on $\bar{F}_N$ carries over to $\tilde{F}_N$, since they differ only by a constant.

Since $|S_1|/|S_0| = \cdots = |S_M|/|S_{M-1}| = 2^n$ and $(\tilde{F}_N)_{S_0} = 0$, then (33) and (6) imply that $|(F_N)_{S_M}| \leq M2^n \|f\|_{\mathrm{BMO}}$. Insert this back into (32) to obtain the uniform quadratic bound

$$\frac{1}{|S_M|}\int_{S_M} |\tilde{F}_N|^2 \leq C'\|f\|^2_{\mathrm{BMO}} < \infty \qquad (N = M, M+1, M+2, \dots).$$

An analogous estimate is also valid for $\{\tilde{G}_N : N \geq M\}$.

For each M, the sequences $\{\tilde{F}_N : N \geq M\}$ and $\{\tilde{G}_N : N \geq M\}$ are consequently bounded in $L^2(S_M)$. By the lemma, the sequence $\{\tilde{R}_N : N \geq M\}$ is also uniformly bounded and equicontinuous on S_M. It is therefore possible to choose a subsequence $N_k \to \infty$, so that $\tilde{F}_{N_k} \rightharpoonup F$ and $\tilde{G}_{N_k} \rightharpoonup G$ weakly in $L^2(S_M)$ and so that $\tilde{R}_{N_k} \to R$ uniformly on S_M.[13] A diagonal argument ensures (with the help of a further subsequence, if necessary) that this convergence holds simultaneously for all M. Since, by (31), $f - f_{S_0} - \tilde{R}_{N_k} = \tilde{F}_{N_k} - \tilde{G}_{N_k}$, then the sequence $\{\tilde{F}_{N_k} - \tilde{G}_{N_k}\}_{N_k}$ must also converge in $L^\infty(S_M)$ to the difference $F - G$.[14] In the limit, (31) thus becomes the asserted decomposition

$$f(x) - f_{S_0} = F(x) - G(x) + R(x) \qquad (x \in \mathbb{R}^n).$$

Note that by construction, R is bounded and uniformly continuous on all of $\mathbb{R}^n$.

To see that $F \in \mathrm{VLO}(\mathbb{R}^n)$, fix an arbitrary cube Q in $\mathbb{R}^n$ and choose M so large that $Q \subseteq S_M$. On this cube S_M, the weak convergence described above implies that there is a sequence $\{\varphi_K\}_{K\in\mathbb{N}}$ of finite convex combinations of the $\tilde{F}_{N_k}$, i.e.,

$$\varphi_K = \sum_{k=1}^{K} t_k \tilde{F}_{N_k} \qquad (t_k \geq 0, \sum_{k=1}^{K} t_k = 1), \tag{34}$$

that converges to F both in $L^2(S_M)$ and (taking a further subsequence, if necessary) pointwise a.e.[15] Now apply Fatou's lemma[16] to this new sequence $\{\varphi_K\}_{K\in\mathbb{N}}$ to obtain from (30) the following bound on the lower mean oscillation of F over Q:

$$\begin{aligned}\frac{1}{|Q|}\int_Q F &\leq \underline{\lim}_K \frac{1}{|Q|}\int_Q \varphi_K = \underline{\lim}_K \sum_{k=1}^{K} \frac{t_k}{|Q|}\int_Q \tilde{F}_{N_k} \\ &\leq \underline{\lim}_K \sum_{k=1}^{K} t_k(\inf_Q \tilde{F}_{N_k} + C_Q) \leq \underline{\lim}_K \sum_{k=1}^{K}(t_k \inf_Q \tilde{F}_{N_k}) + C_Q \\ &\leq \underline{\lim}_K \inf_Q \sum_{k=1}^{K} t_k \tilde{F}_{N_k} + C_Q = \underline{\lim}_K \inf_Q \varphi_K + C_Q \leq \inf_Q F + C_Q.\end{aligned}$$

---

[13] The John-Nirenberg inequality has been invoked to move from uniform boundedness in L^1 to that in L^2; otherwise, weak compactness would have only guaranteed the existence of subsequences $\{\tilde{F}_{N_k}\}$ and $\{\tilde{G}_{N_k}\}$ converging to measures.

[14] It is not claimed—*separately*—that $\tilde{F}_{N_k}(x) \to F(x)$ and $\tilde{G}_{N_k}(x) \to G(x)$ for a.e. x.

[15] See Theorem 3.13 in [12] or Theorem V.1.2 in [16]; in the latter work, this result is attributed to S. Mazur. Note that the coefficients t_k $(1 \leq k \leq K)$ of φ_K in (34) may depend on K.

[16] Suppose that $\{\varphi_K\}$ is a sequence of non-negative, measurable functions that converges a.e. to φ. What is needed here is the (standard) L^1 form of Fatou's lemma, $\int \varphi \leq \underline{\lim}_K \int \varphi_K$, as well as its L^∞ form: $\underline{\lim}_K(\inf \varphi_K) \leq \inf \varphi$; the latter can be verified via a simple proof by contradiction. Recall that we write $\inf \varphi$ for ess $\inf \varphi$, as indicated in the introduction.

As, by assumption, $\sup_{Q\subseteq S_M} C_Q < \infty$ and $\lim_{\ell(Q)\to 0} C_Q = 0$, then $F \in \mathrm{VLO}(S_M)$ for each M. The treatment of G is identical. Modulo the proof of Lemma 3, the proof of the theorem is now complete.

Proof of Lemma 3. We use the averaging procedure of [6] to move from the dyadic version of the theorem (Lemma 2) to the general, local version (Lemma 3). As above (but now over all of $\mathbb{R}^n$), assume that $\|f\|_{\mathrm{BMO}} = 2^{-n}$. For each $\alpha \in \mathbb{R}^n$, let $T_\alpha f$ denote the translate of f by α, where $T_\alpha f(x) = f(x-\alpha)$. Choose a single, strictly decreasing sequence $\{\delta_j\}_{j\in\mathbb{N}} \subset 2^{-\mathbb{N}}$ of dyadic scales so that (19) holds uniformly for $T_\alpha f$ (in place of f) as α varies over $\mathbb{R}^n$; this is possible due to the assumption that $f \in \mathrm{VMO} = \mathrm{VMO}(\mathbb{R}^n)$.

Fix N and assume, without loss of generality, that $f_{S_N} = 0$. Set $Q_0 = S_{N+1}$ and $\delta_0 = \ell(S_{N+1})$. For each $\alpha \in S_N$, apply Lemma 2 to $T_\alpha f$ on Q_0. The result is that

$$T_\alpha f(x) - (T_\alpha f)_{S_{N+1}} = F^{(\alpha)}(x) - G^{(\alpha)}(x) + R^{(\alpha)}(x) \qquad (x \in S_{N+1}),$$

where $F^{(\alpha)}$, $G^{(\alpha)}$, and $R^{(\alpha)}$ are defined as in (21). In particular, $F^{(\alpha)} = \sum_{j=1}^{\infty} F_j^{(\alpha)}$, where $F_j^{(\alpha)}$ is locally constant on the mesh of dyadic subcubes of Q_0 of edge-length δ_j. In addition, the expansion (25) guarantees that there are non-negative coefficient functions $a_k^{(\alpha)}$, depending measurably on α, such that

$$F^{(\alpha)}(x) = \sum_{Q_k\in\mathcal{D}(S_{N+1})} a_k^{(\alpha)} \chi_{Q_k}(x).$$

Note that this sum runs over $\mathcal{D}(S_{N+1})$, a fixed, countable collection of cubes that is indexed by k and independent of α; as discussed in the remark at the end of §2, when $\delta_j < \ell(Q_k) \le \delta_{j-1}$, each coefficient $a_k^{(\alpha)}$ is either 0 or in the interval $(2^{1-j}, 2^{2-j}]$. Equation (26) leads to a similar representation for $G^{(\alpha)}$, and (27) gives rise to the sum

$$R^{(\alpha)}(x) = \sum_{Q_k\in\mathcal{D}(S_{N+1})} c_k^{(\alpha)} \chi_{Q_k}(x); \tag{35}$$

when $\ell(Q_k) = \delta_j$ for some $j \in \mathbb{N}$, then $|c_k^{(\alpha)}| \le 2^{2-j}$, and $c_k^{(\alpha)} = 0$ otherwise.

Since $f(x) = (1/|S_N|)\int_{S_N} f(x)\,d\alpha$, then for a.e. $x \in S_N$

$$\begin{aligned} f(x) &= \frac{1}{|S_N|}\int_{S_N} T_{-\alpha}(T_\alpha f)(x)\,d\alpha \\ &= \frac{1}{|S_N|}\int_{S_N} T_{-\alpha}\big(F^{(\alpha)} - G^{(\alpha)} + R^{(\alpha)} + (T_\alpha f)_{S_{N+1}}\big)(x)\,d\alpha \\ &= \bar{F}_N(x) - \bar{G}_N(x) + \bar{R}_N(x); \end{aligned}$$

in the last line,

$$\bar{F}_N(x) = \frac{1}{|S_N|}\int_{S_N} T_{-\alpha}(F^{(\alpha)})(x)\,d\alpha = \frac{1}{|S_N|}\int_{S_N} F^{(\alpha)}(x+\alpha)\,d\alpha,$$

$\bar{G}_N(x)$ is defined analogously, and

$$\begin{aligned} \bar{R}_N(x) &= \frac{1}{|S_N|}\int_{S_N} T_{-\alpha}\big(R^{(\alpha)}\,d\alpha + (T_\alpha f)_{S_{N+1}}\big)(x)\,d\alpha \\ &= \frac{1}{|S_N|}\int_{S_N} R^{(\alpha)}(x+\alpha) + \frac{1}{|S_N|}\int_{S_N} (T_\alpha f)_{S_{N+1}}\,d\alpha. \end{aligned} \tag{36}$$

It remains to show that these functions $\bar{F}_N$ and $\bar{G}_N$ are in VLO on the cube S_N and that the function $\bar{R}_N$ is uniformly continuous there. In particular, we wish to show that

$$\frac{1}{|Q|}\int_Q \bar{F}_N \le \inf_Q \bar{F}_N + o(1) \qquad \big(\ell(Q) \to 0\big). \tag{37}$$

and that

$$\sup_Q \bar{R}_N \le \inf_Q \bar{R}_N + o(1) \qquad \big(\ell(Q) \to 0\big). \tag{38}$$

To reach this goal, fix an arbitrary cube Q within S_N, and find J such that $\ell(Q) \le \delta_J/(2\sqrt{n})$. Split the terms comprising $F^{(\alpha)}$ according to the size of Q by writing $F^{(\alpha)} = F^{(\alpha)}_{\text{large}} + F^{(\alpha)}_{\text{small}} = \sum_{j=1}^{J} F_j^{(\alpha)} + \sum_{j=J+1}^{\infty} F_j^{(\alpha)}$. Then

$$F^{(\alpha)}_{\text{large}}(x) = \sum_{\ell(Q_k)>\delta_J} a_k^{(\alpha)} \chi_{Q_k}(x), \qquad F^{(\alpha)}_{\text{small}}(x) = \sum_{\ell(Q_k)\le\delta_J} a_k^{(\alpha)} \chi_{Q_k}(x). \tag{39}$$

Note that only finitely many coefficients $a_k^{(\alpha)}$ enter into the first sum in (39), since this sum runs only over those dyadic subcubes $Q_k \in \mathcal{D}(S_{N+1})$ of size greater than δ_J (and this includes only cubes significantly larger than Q).

The corresponding averaged forms are given by

$$\bar{F}_{\text{large}}(x) = \frac{1}{|S_N|}\int_{S_N} F^{(\alpha)}_{\text{large}}(x+\alpha)\,d\alpha, \qquad \bar{F}_{\text{small}}(x) = \frac{1}{|S_N|}\int_{S_N} F^{(\alpha)}_{\text{small}}(x+\alpha)\,d\alpha;$$

thus, $\bar{F}_N = \bar{F}_{\text{large}} + \bar{F}_{\text{small}}$. To prove (37) we shall verify the bounds

$$\sup_Q \bar{F}_{\text{large}} \le \inf_Q \bar{F}_{\text{large}} + CJ2^{-J} \tag{40}$$

and

$$\frac{1}{|Q|}\int_Q \bar{F}_{\text{small}} \le \inf_Q \bar{F}_{\text{small}} + C2^{-J}. \tag{41}$$

Now, the first of these is a consequence of the following Lipschitz estimate[17] on the contribution to $\bar{F}_N$ of the terms associated to cubes of a similar size. For this, let

$$\hat{F}_j(x) = \frac{1}{|S_N|}\int_{S_N} \sum_{\delta_j<\ell(Q_k)\le\delta_{j-1}} a_k^{(\alpha)} \chi_{Q_k}(x+\alpha)\,d\alpha,$$

so that $\bar{F}_{\text{large}}(x) = \sum_{j=1}^{J} \hat{F}_j(x)$.

Lemma 4. *If* $\max_{1\le i\le n} |x_i - y_i| \le \delta_j$, *then*

$$|\hat{F}_j(x) - \hat{F}_j(y)| \le C2^{-j}\frac{|x-y|}{\delta_j},$$

with C *dependent only on the dimension* n *(and, in particular, not on* j*).*

[17] This is Lemma 3.2 in [6], re-interpreted to highlight only certain scales.

Proof. Fix $x, y \in S_N$ with $\max_{1\le i\le n} |x_i - y_i| \le \delta_j$. When $\ell(Q_k) \le \delta_{j-1}$, then the corresponding coefficient $a_k^{(\alpha)}$ is no more than 2^{2-j}, as discussed above. So,

$$|\hat{F}_j(x) - \hat{F}_j(y)| \le \frac{1}{|S_N|}\int_{S_N} \sum_{\delta_j<\ell(Q_k)\le\delta_{j-1}} |a_k^{(\alpha)}||\chi_{Q_k}(x+\alpha) - \chi_{Q_k}(y+\alpha)|\,d\alpha$$

$$\le C2^{-j}\sum_{r=0}^{r_j} \frac{1}{|S_N|}\int_{S_N} \sum_{\ell(Q_k)=2^r\delta_j} |\chi_{Q_k}(x+\alpha) - \chi_{Q_k}(y+\alpha)|\,d\alpha,$$

where $r_j = \log_2(\delta_{j-1}/\delta_j)$. The last integrand is twice the characteristic function of the set of all $\alpha \in S_N$ for which $x+\alpha$ and $y+\alpha$ lie in different dyadic subcubes Q_k of S_{N+1} of size $\ell(Q_k) = 2^r\delta_j$. The relative density of this set in S_N is majorized by a constant multiple of $\sum_{i=1}^n |x_i - y_i|/(2^r\delta_j)$. Thus,

$$|\hat{F}_j(x) - \hat{F}_j(y)| \le C2^{-j}\sum_{r=0}^{r_j}\frac{|x-y|}{2^r\delta_j} \le C'2^{-j}\frac{|x-y|}{\delta_j},$$

as claimed. □

Now, suppose that $x, y \in Q \subseteq S_N$ with $\ell(Q) \le \delta_J/(2\sqrt{n})$, as above. Then $|x-y| \le \sqrt{n}\ell(Q) \le \delta_J < \delta_{J-1} < \cdots < \delta_1$. The lemma then gives

$$|\bar{F}_{\text{large}}(x) - \bar{F}_{\text{large}}(y)| \le \sum_{j=1}^{J} |\hat{F}_j(x) - \hat{F}_j(y)| \le C\sum_{j=1}^{J} 2^{-j}\frac{\delta_J}{\delta_j}.$$

Let σ_J denote this last sum. Since $\sigma_1 = 1/2$ and

$$\sigma_{J+1} = \frac{\delta_{J+1}}{\delta_J}\sigma_J + \frac{1}{2^{J+1}} \le \frac{1}{2}\left(\sigma_J + \frac{1}{2^J}\right),$$

then an elementary induction argument shows that $\sigma_J \le J/2^J$. This confirms (40), the bound on the modulus of continuity of $\bar{F}_{\text{large}}$.

What about (41)? The right-hand side there can in fact be simplified further by noting that $\bar{F}_{\text{small}} \ge 0$. As for the left-hand side, from Fubini's theorem it follows that

$$\frac{1}{|Q|}\int_Q \bar{F}_{\text{small}} = \frac{1}{|S_N|}\int_{S_N}\frac{1}{|Q|}\int_{Q+\alpha} F_{\text{small}}^{(\alpha)}(x)\,dx\,d\alpha.$$

For the proof of (41) and hence of (37), it thus suffices to obtain a suitable estimate on the inner integral here, i.e., to show that

$$\frac{1}{|Q|}\int_{Q+\alpha} F_{\text{small}}^{(\alpha)}(x)\,dx = \sum_{j=J+1}^{\infty}\frac{1}{|Q|}\int_{Q+\alpha} F_j^{(\alpha)}(x)\,dx \le C2^{-J} \tag{42}$$

uniformly for all $\alpha \in S_N$. But the translated cube $Q+\alpha$ serving as the region of integration is contained within a union of 2^n congruent dyadic subcubes of S_{N+1}, each having edge-length less than twice that of Q (and hence not more than δ_J). Applying (24) in §2 to each of these subcubes and summing up leads to the bound $\int_{Q+\alpha} F_j^{(\alpha)}(x)\,dx \le C2^{-j}|Q|$. A further sum in j (for $j \ge J+1$) leads to (42). The proof of estimate (37) for $\bar{G}_N$ is similar.

It remains to show that $\bar{R}_N$, as defined in (36), is uniformly continuous on S_N. The second term in (36), the average of $(T_\alpha f)_{S_{N+1}}$ over $\alpha \in S_N$, is constant and can safely be ignored. In addition, the countable sum giving $R^{(\alpha)}$ in (35) can be split according to the size of Q, so that $R^{(\alpha)} = R^{(\alpha)}_{\text{large}} + R^{(\alpha)}_{\text{small}}$, as in (39). The partition $\bar{R}_N = \bar{R}_{\text{large}} + \bar{R}_{\text{small}}$ likewise applies to the corresponding averaged forms, up to the constant just described. Since $|c_k^{(\alpha)}| \le C2^{-j}$ when $\ell(Q) \le \delta_j$ (as was the case for the coefficients $a_k^{(\alpha)}$), then the argument in Lemma 4 for $\bar{F}_{\text{large}}$ carries over to $\bar{R}_{\text{large}}$, so that estimate (40) also holds for the latter. On the other hand, since only relatively few of the coefficients $c_k^{(\alpha)}$ may be non-zero (namely those corresponding to cubes Q_k of size exactly δ_j, for some $j \in \mathbb{N}$), the BLO estimate (41) can be replaced by a stronger estimate on the modulus of continuity of $\bar{R}_{\text{small}}$. In fact,

$$\sup_Q |\bar{R}_{\text{small}}| \le C \sum_{j=J}^{\infty} 2^{-j} \le C2^{-J},$$

by (20). All together, then, $\sup_Q \bar{R}_N - \inf_Q \bar{R}_N \le CJ2^{-J}$, when $\ell(Q) \le \delta_J/(2\sqrt{n})$, giving (38). Thus, $\bar{R}_N$ can be re-defined on a set of measure zero to yield a uniformly continuous function.

This settles the last remaining step in the proof of the lemma, and the decomposition theorem is therefore complete. □

References

[1] F. Chiarenza, M. Frasca, and P. Longo, *$W^{2,p}$-solvability of the Dirichlet problem for non-divergence elliptic equations with* VMO *coefficients*, Trans. Amer. Math. Soc. **336** (1993), 841–853.

[2] M. Carozza and R. Mellone, *The distance to L^∞ or to* VMO *for* BMO *functions*, Ricerce Mat. **44** (1995), 439–448.

[3] R. R. Coifman, R. Rochberg, *Another characterization of* BMO, Proc. Amer. Math. Soc. **79** (1980), 249–254.

[4] J. García-Cuerva and J. L. Rubio de Francia, *Weighted norm inequalities and related topics*, North-Holland, Amsterdam, New York, and Oxford, 1985.

[5] J. Garnett and P. Jones, *The distance in* BMO *to L^∞*, Ann. of Math. **108** (1978), 373–393.

[6] J. Garnett and P. Jones, BMO *from dyadic* BMO, Pacific J. Math. **99** (1982), 351–371.

[7] P. J. Holden, *Extension theorems for functions of vanishing mean oscillation*, Pacific J. Math. **142** (1990), 277–295.

[8] F. John and L. Nirenberg, *On functions of bounded mean oscillation*, Comm. Pure Appl. Math. **14** (1961), 415–426.

[9] P. Jones, *Factorization of A_p weights*, Ann. of Math. **111** (1980), 511–530.

[10] C. E. Kenig and T. Toro, *Free boundary regularity for harmonic measures and Poisson kernels*, Ann. of Math. by 150 (1999), 369–454.

[11] M. B. Korey, *Optimal factorization of Muckenhoupt weights*, Trans. Amer. Math. Soc. **353** (2001), 839–851.

[12] W. Rudin, *Functional analysis*, McGraw-Hill, New York, 1973.

[13] D. Sarason, *Functions of vanishing mean oscillation*, Trans. Amer. Math. Soc. **207** (1975), 391–405.

[14] X. L. Shi and A. Torchinsky, *Functions of vanishing mean oscillation*, Math. Nachr. **133** (1987), 289–296.

[15] S. Spanne, *Some function spaces defined using the mean oscillation over cubes*, Ann. Scuola Norm. Sup. Pisa **19** (1965), 593–608.

[16] K. Yosida, *Functional analysis* Grundlehren Math. Wiss., vol. 123, Springer-Verlag, Berlin, Heidelberg, and New York, 1965.

INSTITUT FÜR MATHEMATIK, UNIVERSITÄT POTSDAM, 14415 POTSDAM, GERMANY
E-mail address: mike@math.uni-potsdam.de

Contemporary Mathematics
Volume **277**, 2001

General Second Order, Strongly Elliptic Systems in Low Dimensional Nonsmooth Manifolds

Dorina Mitrea and Marius Mitrea

1. Introduction

Let Ω be a Lipschitz domain and consider the Dirichlet Poisson problem for a second-order, elliptic differential operator L, i.e.

$$Lu = f \quad \text{in} \quad \Omega, \qquad u\big|_{\partial\Omega} = g \quad \text{on} \quad \partial\Omega. \tag{1.1}$$

The situation when $g \in L^p(\partial\Omega)$, $f = 0$ and nontangential maximal function estimates are sought for the solution has received a lot of attention. For example, [**9**], [**19**], [**24**], [**50**], [**11**] contain a complete theory of the Euclidean, flat-space Laplacian. In the same geometric context, much progress has been made in the treatment of strongly elliptic systems; cf. [**13**], [**18**], [**12**], [**17**], [**34**]. There are also parabolic and higher degree versions of these constant coefficient results; see [**3**], [**41**], [**44**], [**4**], [**23**]. An excellent survey of the state of the art in this active field of research up to the early 90's can be found in [**28**].

More recently, starting with the pioneering work in [**25**], the scope of these investigations has been extended to include the case when the data f, g (and, hence, the solution u) are in *Sobolev-Besov* spaces. In [**25**], Jerison and Kenig have been able to use harmonic measure techniques in order to obtain sharp results for the Euclidean Laplacian with Dirichlet boundary conditions. Subsequently, a new approach has been developed in [**20**] where the authors have succeeded to prove sharp invertibility results for the classical layer potential operators on Sobolev-Besov scales. In particular, in [**20**] an alternative solution of the main results in [**25**] was given, which can handle Neumann boundary conditions as well. In principle, this method is also applicable to the case of systems of equations; the case of the three dimensional, constant coefficient Lamé system has been dealt with in [**31**].

In [**35**], [**36**], [**37**], [**38**], [**33**], a program aimed at extending these types of results to the case of variable coefficient PDE's (and, more generally, PDE's on

2000 *Mathematics Subject Classification.* Primary 35J25, 46E35; Secondary 42B20, 45B05.

Key words and phrases. Layer potentials, Besov spaces, Sobolev spaces, atomic Hardy spaces, Poisson problems, elliptic systems, Lipschitz domains.

The first author was supported in part by a University of Missouri Research Board Grant and a Summer Research Fellowship. The second author was supported in part by NSF.

manifolds) has been initiated. The goal of the present paper is to continue this line of research by considering (1.1) for arbitrary, second order, strongly elliptic, formally self-adjoint, variable coefficient systems L in two and three-dimensional Lipschitz domains.

Working on Sobolev-Besov scales is both natural and has its distinct advantages. For example, it allows for an all-encompassing, unified treatment of several (otherwise not directly related) classical function spaces (such as Hölder, L^p and Hardy classes), and provides a context where the method of layer potentials works well. In recent years, such results have found important applications in nonlinear PDE's ([**43**], [**16**]) and nonlinear approximation ([**14**], [**15**]); cf. also [**25**], [**30**] for other ramifications.

Our strategy is to interpolate between various end-point results. One illustrative instance of the latter, occuring in connection with the two-dimensional Lamé operator in Lipschitz domains is as follows. Let p_{Dir} be the critical exponent for the Dirichlet problem on L^p spaces, i.e. the infimum of all p's such that

$$\mu\Delta u + (\lambda+\mu)\mathrm{grad}\,\mathrm{div}\,u = 0 \text{ in } \Omega, \quad u\big|_{\partial\Omega} = f \text{ on } \partial\Omega, \tag{1.2}$$

is well-posed for boundary data in $L^p(\partial\Omega)$ (when a nontangential maximal function estimate is sought for u). Then (1.2) is also well-posed for data in $C^\alpha(\partial\Omega)$, and when the solution u is looked for in the class $C^\alpha(\bar\Omega)$, if $0 < \alpha < 1/p_{Dir}$. Note that, generally speaking, $p_{Dir} < 2$ so that $1/p_{Dir} > 1/2$. Moreover, the solution can be represented in the form of a single layer and natural estimates hold. In fact, we show that the same type of result is valid for much more general (variable coefficient) systems.

The organization of the paper is as follows. In §2 we introduce notation and collect basic results. In §3 we derive *a priori* estimates on atomic Hardy spaces and derive invertibility results. Our first major result, Theorem 4.2 contained in §4, describes the sharp invertibility region for the single layer potential operator on Besov classes; cf. also Theorem 4.1 for Sobolev scales. The main technical novelty is the observation that the atomic H^p-theory, originally discovered in [**11**] when $p = 1$ in all space dimensions (cf. also [**12**]), actually extends to $p = \frac{2}{3}$ in dimension two. With a different finality in mind, this has been also pointed out in [**32**].

Applications are discussed in §5, where most of our main results are collected; see Theorem 5.1, Theorem 5.2, Theorem 5.3, Theorem 5.4, Theorem 5.5, and Theorem 5.7. These deal, respectively, with the Dirichlet Poisson problem on Sobolev-Besov spaces for L, the L^p-Dirichlet and regularity problems for L in Dahlberg's sense, fractional powers of L, square-root and Green function estimates, and invertibility results for layer potentials associated with the two-dimensional Laplace-Beltrami operator. Finally, further research is outlined in §6, where the case $\dim M = 3$ is emphasized. The two dimensional case is somewhat special both in terms of the specific form of the results valid in this context and in terms of the techniques employed. We would also like to mention that, at least in the case of systems, the approach employed here is essentially restricted to dimensions ≤ 3, and that dealing with similar problems for higher dimensional systems remains very much an open problem at the moment.

The methods and results of this paper owe a great deal to earlier work of many people. The authors would like to take this opportunity to express their gratitude to Professor Carlos Kenig and Professor Michael Taylor for their encouragement and support. We also thank the referee for several suggestions which led to an overall improvement in the presentation.

2. Notation, definitions and preliminary results

Throughout this paper we shall assume that M is a smooth, compact, boundaryless Riemannian manifold of (real) dimension ≤ 3. Sections §2-§5 deal with the case when $\dim M = 2$. In §6, we consider the case $\dim M = 3$. Recall that a subdomain of M is called *Lipschitz* provided its boundary can be described, in appropriate local coordinates, by means of graphs of Lipschitz functions.

For a fixed, connected Lipschitz domain Ω in M, we denote by $d\sigma$ the surface measure on $\partial\Omega$ and by ν the outward unit conormal defined a.e. on the boundary. Throughout the paper, $C = C(\partial\Omega)$ will denote various constants which depend exclusively on the Lipschitz character of Ω. Next, for some large $\kappa_o > 0$ and each $x \in \partial\Omega$ we set

$$\gamma(x) := \{y \in \Omega;\ \mathrm{dist}\,(y,x) \leq \kappa_o\, \mathrm{dist}\,(y,\partial\Omega)\}. \tag{2.1}$$

Then if u is defined in Ω, $\mathcal{N}(u)$, the nontangential maximal function of u, is defined at boundary points by

$$\mathcal{N}(u)(x) := \sup\{|u(y)| : y \in \gamma(x)\}, \qquad x \in \partial\Omega. \tag{2.2}$$

For $1 < p < \infty$, we denote by L^p the space of p-integrable functions (which will be defined either over Ω or $\partial\Omega$), and by $L^p_1(\partial\Omega)$ the Sobolev space of functions in $L^p(\partial\Omega)$ with tangential gradients in $L^p(\partial\Omega)$. Spaces with fractional indices of smoothness can be obtained by complex interpolation, i.e. $[L^p(\partial\Omega), L^p_1(\partial\Omega)]_\theta = L^p_\theta(\partial\Omega)$, $0 \leq \theta \leq 1$, $1 < p < \infty$. We also set $L^p_{-s}(\partial\Omega) := (L^{p'}_s(\partial\Omega))^*$, for $1 < p, p' < \infty$ conjugate exponents, $0 \leq s \leq 1$.

For each $0 < |s| < 1$, $1 \leq p, q \leq \infty$, we let $B^{p,q}_s(\partial\Omega)$ stand for the usual scale of Besov spaces over $\partial\Omega$ with smoothness s.

We recall that, if Ω is a Lipschitz domain in $\mathbb{R}^2$, then, for $1 \leq p, q < \infty$ and $0 < s < 1$, one definition of the Besov space $B^{p,q}_s(\partial\Omega)$ is the collection of all measurable functions f on $\partial\Omega$ such that

$$\|f\|_{B^{p,q}_s(\partial\Omega)} := \|f\|_{L^p(\partial\Omega)} + \left(\int_{\partial\Omega}\Big(\int_{\partial\Omega} \frac{|f(x)-f(y)|^p}{|x-y|^{(1+sq)p/q}}\, d\sigma(y)\Big)^{q/p} d\sigma(x)\right)^{1/q} < \infty. \tag{2.3}$$

The case $0 < s < 1$, $p = q = \infty$ corresponds to the non-homogeneous version of the space of Hölder continuous functions on $\partial\Omega$. That is, $f \in B^{\infty,\infty}_s(\partial\Omega)$, if

$$\|f\|_{B^{\infty,\infty}_s(\partial\Omega)} := \|f\|_{L^\infty(\partial\Omega)} + \sup_{x \neq y \in \partial\Omega} \frac{|f(x)-f(y)|}{|x-y|^s} < \infty. \tag{2.4}$$

Next, we consider the case when Ω is a Lipschitz domain in the compact Riemannian manifold M. It is then natural to say that f belongs to $B^{p,q}_s(\partial\Omega)$ for some $1 \leq p, q \leq \infty$ and $0 < s < 1$ if (and only if) for any smooth chart $(\mathcal{O}, \Phi)$ and any smooth cut-off function $\theta \in C^\infty_{\mathrm{comp}}(\mathcal{O})$, $(f\theta) \circ \Phi^{-1} \in B^{p,q}_s(\Phi(\mathcal{O} \cap \partial\Omega))$. It is then elementary to transfer basic results about Besov spaces originally proved in the classical Euclidean setting to the case of boundaries of Lipschitz domains in Riemannian manifolds. For example, real interpolation gives that $(L^p(\partial\Omega), L^p_1(\partial\Omega))_{\theta,q} = B^{p,q}_\theta(\partial\Omega)$, with $0<\theta<1$, $1<p,q<\infty$. Also, for $0<\theta<1$, $1\leq p_0,p_1,q_0,q_1\leq\infty$ and $0<s_0,s_1<1$, there holds

$$[B^{p_0,q_0}_{s_0}(\partial\Omega), B^{p_1,q_1}_{s_1}(\partial\Omega)]_\theta = B^{p,q}_s(\partial\Omega) \tag{2.5}$$

where $\frac{1}{p} := \frac{1-\theta}{p_0} + \frac{\theta}{p_1}$, $\frac{1}{q} := \frac{1-\theta}{q_0} + \frac{\theta}{q_1}$ and $s := (1-\theta)s_0 + \theta s_1$. A similar result is valid for the real method of interpolation.

As usual, Besov spaces with negative indices of smoothness are defined via duality. Specifically, $B^{p,q}_{-s}(\partial\Omega) := \left(B^{p',q'}_s(\partial\Omega)\right)^*$ for each $0<s<1$, $1<p,q<\infty$ and $p' = (1-\frac{1}{p})^{-1}$, $q' = (1-\frac{1}{q})^{-1}$.

In the sequel, we shall also need to work with the Besov spaces $B^{1,1}_{-s}(\partial\Omega)$, $s\in(0,1)$. Inspired by the corresponding atomic characterization from [**22**], set

$$B^{1,1}_{-s}(\partial\Omega) := \mathcal{C} + \Big\{f = \sum_{j\geq 0}\lambda_j a_j : a_j\ B^{1,1}_{-s}(\partial\Omega) - \text{atom},\ (\lambda_j)_j \in \ell^1\Big\}, \tag{2.6}$$

where the series converges in the sense of distributions, and $\mathcal{C}$ is the space of constant functions on $\partial\Omega$. In this two-dimensional context, a $B^{1,1}_{-s}(\partial\Omega)$-atom, $0<s<1$, is a function $a\in L^\infty(\partial\Omega)$ with support contained in $B_r(x_0)\cap\partial\Omega$ for some $x_0\in\partial\Omega$, $0<r<\operatorname{diam}\Omega$, and satisfying

$$\int_{\partial\Omega} a\,d\sigma = 0, \qquad \|a\|_{L^\infty(\partial\Omega)} \leq r^{-s-1}. \tag{2.7}$$

Furthermore, for $f\in B^{1,1}_{-s}(\partial\Omega)$, $0<s<1$,

$$\|f\|_{B^1_{-s}(\partial\Omega)} := \inf\Big\{\|g\|_{L^\infty} + \sum_{j\geq 0}|\lambda_j| : f = g + \sum_{j\geq 0}\lambda_j a_j\Big\} \tag{2.8}$$

where $g\in\mathcal{C}$, a_j's and $(\lambda_j)_j$ are as in (2.6).

We now briefly discuss the case of Besov and Sobolev classes in the interior of a Lipschitz domain $\Omega\subset M$. First, for $1\leq p,q\leq\infty$, $s>0$, the Besov space $B^{p,q}_s(M)$ is defined by localizing and transporting via local charts its Euclidean counterpart, i.e., $B^{p,q}_s(\mathbb{R}^n)$ (for the latter see, e.g., [**40**], [**2**], [**1**], [**49**], [**26**]). Going further, $B^{p,q}_s(\Omega)$ consists of restrictions to Ω of functions from $B^{p,q}_s(M)$. This is equipped with the natural norm, i.e., defined by taking the infimum of the $\|\cdot\|_{B^{p,q}_s(M)}$-norms of all possible extensions to M.

Using Stein's extension operator and then invoking well known real interpolation results (cf., e.g., [**2**]), it follows that for any Lipschitz domain $\Omega\subset M$,

$$(B^{p_0,q_0}_{s_0}(\Omega), B^{p_1,q_1}_{s_1}(\Omega))_{\theta,p} = B^{p,q}_s(\Omega) \tag{2.9}$$

if $\frac{1}{p} = \frac{1-\theta}{p_0} + \frac{\theta}{p_1}$, $\frac{1}{q} = \frac{1-\theta}{q_0} + \frac{\theta}{q_1}$, $s = (1-\theta)s_0 + \theta s_1$, $0<\theta<1$, $1\leq p_0,p_1,q_0,q_1\leq\infty$, $s_0\neq s_1$, $s_0,s_1>0$.

A similar discussion applies to the Sobolev (or potential) spaces $L^p_s(\Omega)$, this time starting with the potential spaces $L^p_s(M)$ (lifted to M from $\mathbb{R}^2$; for the latter context see, e.g., [**45**], [**2**]). Following [**25**], for $s \in \mathbb{R}$ we define the space $L^p_{s,0}(\Omega)$ to consist of distributions in $L^p_s(M)$ supported in $\overline{\Omega}$ (with the norm inherited from $L^p_s(M)$). It is known that $C^\infty_{\text{comp}}(\Omega)$ is dense in $L^p_{s,0}(\Omega)$ for all values of s and p.

Recall (cf. [**26**]) that the trace operator

$$\text{Tr} : L^p_s(\Omega) \longrightarrow B^{p,p}_{s-\frac{1}{p}}(\partial\Omega) \tag{2.10}$$

is well defined, bounded and onto if $1 < p < \infty$ and $\frac{1}{p} < s < 1 + \frac{1}{p}$. This also has a bounded right inverse whose operator norm is controlled exclusively in terms of p, s and the Lipschitz character of Ω. Similar results are valid for $\text{Tr} : B^{p,q}_s(\Omega) \to B^{p,q}_{s-\frac{1}{p}}(\partial\Omega)$. In this latter case we may allow $1 \leq p, q \leq \infty$; cf. [**2**].

Next, if $1 < p < \infty$ and $\frac{1}{p} < s < 1 + \frac{1}{p}$, the space $L^p_{s,0}(\Omega)$ is the kernel of the trace operator Tr acting on $L^p_s(\Omega)$. This follows from the Euclidean result [**25**, Proposition 3.3]. In fact, for the same range of indices, $L^p_{s,0}(\Omega)$ is the closure of $C^\infty_{\text{comp}}(\Omega)$ in the $L^p_s(\Omega)$ norm.

For positive s, $L^p_{-s}(\Omega)$ is defined as the space of linear functionals on test functions in Ω equipped with the norm

$$\|f\|_{L^p_{-s}(\Omega)} := \sup\left\{|\langle f, g\rangle| : g \in C^\infty_{\text{comp}}(\Omega),\ \|\tilde{g}\|_{L^q_s(M)} \leq 1\right\} \tag{2.11}$$

where tilde denotes the extension by zero outside Ω and $\frac{1}{p} + \frac{1}{q} = 1$. For all values of p and s, $C^\infty(\overline{\Omega})$ is dense in $L^p_s(\Omega)$. Also, for any $s \in \mathbb{R}$,

$$L^q_{-s,0}(\Omega) = (L^p_s(\Omega))^* \quad \text{and} \quad L^p_{-s}(\Omega) = \left(L^q_{s,0}(\Omega)\right)^*. \tag{2.12}$$

For later reference, let us point out that $L^p_{\alpha,0}(\Omega) = L^p_\alpha(\Omega)$ for $0 \leq \alpha < \frac{1}{p}$, $1 < p < \infty$ (the latter can be easily deduced from Proposition 3.5 in [**25**]).

We shall also need the fact that the gradient operator

$$\nabla : L^p_s(\Omega) \longrightarrow L^p_{s-1}(\Omega, TM) \tag{2.13}$$

is well defined and bounded for $1 < p < \infty$ and $s > 1/p - 1$. This can be proved much as in [**20**].

Going further, it is well known that $L^p_s(\Omega)$, $L^p_{s,0}(\Omega)$, $L^p_{-s}(\Omega)$, $L^p_{-s,0}(\Omega)$, $L^p_s(\partial\Omega)$ and $L^p_{-s}(\partial\Omega)$ are complex interpolation scales for $1 < p < \infty$ and nonnegative s (in the case of the last two scales we also require that $s \leq 1$). Also, the Besov and Sobolev spaces on the domain are related via real interpolation. For instance, we have the formula

$$(L^p(\Omega), L^p_k(\Omega))_{s,q} = B^{p,q}_{sk}(\Omega) \tag{2.14}$$

when $0 < s < 1$, $1 < p < \infty$ and k is a nonnegative integer. A more detailed discussion and further properties of these spaces, as well as proofs for some of the statements in this paragraph for Euclidean domains can be found in [**2**], [**1**], [**25**].

We continue to recall some definitions. Call $\vartheta \in L^\infty(\partial\Omega)$ a $\mathfrak{H}^p_{\text{at}}(\partial\Omega)$-*atom* (or simply atom), $\frac{1}{2} < p \leq 1$, if $\text{supp}\,\vartheta \subseteq B_r(x_0) \cap \partial\Omega$ for some $x_0 \in \partial\Omega$, $r \in (0, \text{diam}\,\Omega]$, $\int_{\partial\Omega} \vartheta\, d\sigma = 0$ and $\|\vartheta\|_{L^\infty(\partial\Omega)} \leq r^{-1/p}$. The latter normalization can be changed to $\|\vartheta\|_{L^q(\partial\Omega)} \leq r^{-1/p+1/q}$ for some fixed $2 \leq q < \infty$, ultimately yielding the same atomic space. See [**7**]. Then f is said to belong to $\mathfrak{H}^p_{\rm at}(\partial\Omega)$ provided it can be written in the form

$$f=\sum_{j\geq 1}\lambda_j\vartheta_j,\quad \vartheta_j \text{ atom},\quad \sum_{j\geq 1}|\lambda_j|^p<\infty. \tag{2.15}$$

We also introduce

$$\|f\|_{\mathfrak{H}^p_{\rm at}(\partial\Omega)}:=\inf\Big\{\big(\sum|\lambda_j|^p\big)^{1/p};\, f=\sum\lambda_j\vartheta_j,\,\vartheta_j\text{ atom}\Big\}. \tag{2.16}$$

This corresponds to the approach in [**7**] considering $\partial\Omega$ equipped with the measure $d\sigma$ and the geodesic distance as a space of homogeneous type. Then we can set

$$\mathfrak{h}^p_{\rm at}(\partial\Omega):=\mathfrak{H}^p_{\rm at}(\partial\Omega)+L^q(\partial\Omega),\quad \forall\, q\in(1,\infty], \tag{2.17}$$

and equip it with the natural norm. The space $\mathfrak{h}^p_{\rm at}(\partial\Omega)$ is "local" in the sense that, under $f\mapsto\varphi f$, it is a module over $B^{\infty,\infty}_\alpha(\partial\Omega)=C^\alpha(\partial\Omega)$, for any $\alpha>\frac1p-1$. Let us also point out that, if $\frac12<p<1$, then $\mathfrak{h}^p_{\rm at}(\partial\Omega)$ is only a quasi-Banach space and

$$(\mathfrak{h}^p_{\rm at}(\partial\Omega))^*=B^{\infty,\infty}_s(\partial\Omega),\qquad s:=\tfrac1p-1. \tag{2.18}$$

Next, a (regular) $\mathfrak{H}^{1,p}_{\rm at}(\partial\Omega)$-atom, for $\frac12<p\leq 1$, is a function $\vartheta\in{\rm Lip}$ satisfying

$$\operatorname{supp}\vartheta\subseteq B_r(x_0)\cap\partial\Omega\quad\text{and}\quad \|\partial_\tau\vartheta\|_{L^\infty(\partial\Omega)}\leq r^{-1/p} \tag{2.19}$$

for some $x_0\in\partial\Omega$, $r\in(0,\operatorname{diam}\Omega]$, where ∂_τ stands for the tangential derivative along $\partial\Omega$ (recall that we are assuming $\dim\Omega=2$). In particular, $\partial_\tau\vartheta$ is a $\mathfrak{H}^p_{\rm at}(\partial\Omega)$-atom. Then the space $\mathfrak{H}^{1,p}_{\rm at}(\partial\Omega)$ is defined as the ℓ^p-span of (regular) atoms, and is equipped with the natural "norm". Once again, regular atoms can be normalized in $L^q_1(\partial\Omega)$, ultimately yielding the same regular atomic space; see also [**11**] when $p=1$.

A simple yet useful observation (seen more or less directly from definitions) is that

$$\partial_\tau:\mathfrak{H}^{1,p}_{\rm at}(\partial\Omega)\longrightarrow\mathfrak{H}^p_{\rm at}(\partial\Omega) \tag{2.20}$$

is well-defined and bounded (in fact Fredholm) for each $\frac12<p\leq 1$. Another important ingredient in our subsequent analysis is as follows.

Let $\mathcal{E},\mathcal{F}\to M$ be two (smooth, Hermitian) vector bundles over the manifold M and assume that $L:\mathcal{E}\to\mathcal{F}$ is a second order, differential operator. Suppose that in local coordinate patches (over which $\mathcal{E},\mathcal{F}$ are trivial) L is given by

$$Lu=\sum_{j,k}\partial_jA^{jk}(x)\partial_ku+\sum_j B^j(x)\partial_ju-V(x)u. \tag{2.21}$$

Above, $A^{jk}:=\big(a^{\alpha\beta}_{jk}\big)_{\alpha,\beta}$, $B^j:=\big(b^{\alpha\beta}_j\big)_{\alpha,\beta}$ and $V:=\big(v^{\alpha\beta}\big)_{\alpha,\beta}$ are matrix-valued functions with entries satisfying

$$(2.22) \qquad a_{jk}^{\alpha\beta} \in L_2^r, \quad b_j^{\alpha\beta} \in L_1^r, \quad v^{\alpha\beta} \in L^r, \quad \text{for some } r > \dim M,$$

We note that the Laplace-Beltrami operator on scalar functions satisfies (2.22) when the metric tensor is Lipschitz. Also, the Hodge Laplacian on 1-forms, satisfies (2.22) provided the metric tensor satisfies

$$(2.23) \qquad g_{jk} \in L_2^r, \quad \text{some } r > \dim M,$$

We define an *ad-hoc* conormal derivative operator as follows. Let $(U_\mu)_\mu$ be an open cover of $\partial\Omega$ and let $(\psi_\mu)_\mu$ be a smooth partition of unity subordinated to it. If L is written in the form (2.21) in each U_μ with coefficients A_μ^{jk}, B_μ^j, V_μ, then we set

$$(2.24) \qquad \partial_\nu u := \sum_\mu \sum_{j,k} \nu_j \psi_\mu A_\mu^{jk} \partial_k u,$$

where ν_j are the components of the unit conormal.

PROPOSITION 2.1. *Let $u \in C^1_{\rm loc}(\Omega)$ be so that $Lu = 0$ in Ω and $\mathcal{N}(\nabla u) \in L^p(\partial\Omega)$ for some $\frac{1}{2} < p \leq 1$. Then $\partial_\nu u$, $\partial_\tau u \in \mathfrak{h}^p_{\rm at}(\partial\Omega)$ and*

$$(2.25) \qquad \|\partial_\nu u\|_{\mathfrak{h}^p_{\rm at}(\partial\Omega)}, \ \|\partial_\tau u\|_{\mathfrak{h}^p_{\rm at}(\partial\Omega)} \leq C(\partial\Omega, p)\|\mathcal{N}(\nabla u)\|_{L^p(\partial\Omega)}.$$

PROOF. The problem is local in character and, hence, it suffices to verify the conclusions we seek in the case when Ω is contained in a coordinate patch where L can be written in the form (2.21) (and when the cutoff functions ψ_μ are absent in (2.24)).

In this setting, the membership to $\mathfrak{h}^p_{\rm at}(\partial\Omega)$ and (2.25) are consequences of the results in [**51**]. There, the context is that of the upper-half space but since the approach utilizes only nontangential maximal function estimates and cancellations based on integrations by parts, it can be extended to the present setting. Parenthetically, let us also note that, at least when $p = 1$, an alternative approach, discovered by Dahlberg and Kenig [**11**] (and which, in turn, is based on an extension result of N. Varopoulos), works as well. □

Next, we derive a Caccioppoli estimate for two-dimensional, strongly elliptic systems.

PROPOSITION 2.2. *Assume that $L : \mathcal{E} \to \mathcal{E}$ is a formally self-adjoint, strongly elliptic, second order operator locally given by (2.21). Then, if rank$\mathcal{E} = 2$, we have for each $x_0 \in \partial\Omega$ and $0 < R < \operatorname{diam}\Omega$,*

$$(2.26) \qquad \iint_{\Omega\cap B_R(x_0)} |\nabla u|^2 \leq \frac{C_{L,\Omega}}{R^2} \iint_{\Omega\cap B_{2R}(x_0)} |u|^2,$$

uniformly for all $u \in L^2_1(\Omega\cap B_{2R}(x_0))$ with $Lu = 0$ in $\Omega\cap B_{2R}(x_0)$, $u|_{\partial\Omega\cap B_{2R}(x_0)} = 0$.

PROOF. The crux of the matter is to show that the coefficients of the principal part of L can be chosen so that, locally, for all tensors $\zeta = (\zeta_j^\alpha)$

$$(2.27) \qquad \sum_{\alpha,\beta} \sum_{j,k} a_{jk}^{\alpha\beta}(x)\zeta_j^\alpha \zeta_k^\beta \geq C|\zeta|^2, \qquad \text{uniformly in } x \in \bar{\Omega}.$$

Then the usual proof of Caccioppoli's inequality (cf., e.g., [**6**, Theorem 2.1, p. 134] applies and yields (2.26).

With an eye on (2.27), let us introduce $A := (a^{11}_{jk})_{j,k}$, $B := (a^{12}_{jk})_{j,k}$ and $C := (a^{22}_{jk})_{j,k}$ so that

$$(a^{\alpha\beta}_{jk})_{\alpha,\beta,j,k} = \begin{bmatrix} A & B \\ B^t & C \end{bmatrix}. \tag{2.28}$$

The strong ellipticity assumption on L translates into $\sum_{j,k} a^{\alpha\beta}_{jk}(x)\xi_j\xi_k \geq \kappa|\xi|^2$ for some $\kappa > 0$ uniformly in x which, in turn, entails $\lambda^2 A + \lambda(B + B^t) + C \geq \kappa$ for all $\lambda \in \mathbb{R}$.

At this stage we would like to quote a general result to the effect that if A, C are symmetric and $\lambda^2 A + \lambda(B + B^t) + C$ is positive semi-definite for each $\lambda \in \mathbb{R}$ then there exists an anti-symmetric matrix $D = -D^t$ so that

$$\begin{bmatrix} A & B + D \\ B^t - D & C \end{bmatrix} \tag{2.29}$$

is positive semi-definite. When A, B, C have constant entries this may be deduced from [**42**, Theorem 5.51] (a direct proof is also given in [**39**]). The interested reader is also referred to [**21**] which contains a survey of work on necessary and sufficient conditions for a pair of quadratic forms to admit a positive definite linear combination. The case of matrices with variable entries is easily seen from this and a routine partition of unity argument.

Clearly, applying this result to $(a^{\alpha\beta}_{jk}) - \epsilon I$ for $\epsilon > 0$ small it follows that in the case when $1 \leq \alpha, \beta \leq 2$ matters can always be arranged so that (2.27) is true. The proof is finished. □

Next, we introduce the single layer potential operator and record some of its main properties, following work in [**33**], [**37**]. The theorem below is valid in *all* space dimensions.

THEOREM 2.3. *Let $\mathcal{E}, \mathcal{F} \to M$ be two smooth vector bundles over the smooth compact, boundaryless manifold M of real dimension $m \geq 2$. It is assumed that the metric structures on $\mathcal{E}$, $\mathcal{F}$ and M have coefficients in L^2_r for some $r > m =$ $\dim M$. Let $L : \mathcal{E} \to \mathcal{F}$ be an elliptic, second-order differential operator mapping C^2 sections of $\mathcal{E}$ into measurable sections of $\mathcal{F}$. It is assumed that, when written in local coordinates, the coefficients of L and L^* satisfy (2.21). Moreover, suppose that*

$$L : L^2_1(M, \mathcal{E}) \longrightarrow L^2_{-1}(M, \mathcal{F}) \tag{2.30}$$

is invertible, and denote by E the Schwartz kernel of L^{-1}. Finally, let Ω be an arbitrary Lipschitz domain in M and introduce the single layer potential operator

$$\mathcal{S}f(x) := \int_{\partial\Omega} \langle E(x, y), f(y)\rangle \, d\sigma(y), \qquad x \in \Omega, \tag{2.31}$$

as well as its boundary trace

$$(2.32)\qquad Sf(x) := \int_{\partial\Omega} \langle E(x,y), f(y)\rangle \, d\sigma(y), \qquad x \in \partial\Omega.$$

Then, with $p^* := \max\{p, 2\}$,

$$(2.33)\qquad \mathcal{S} : L^p_{-s}(\partial\Omega, \mathcal{F}) \longrightarrow B^{p,p^*}_{-s+1+1/p}(\Omega, \mathcal{E}), \qquad S : L^p_{-s}(\partial\Omega, \mathcal{F}) \longrightarrow L^p_{1-s}(\partial\Omega, \mathcal{E})$$

are bounded operators for each $1 < p < \infty$, $0 \leq s \leq 1$. *Moreover, if* $1 \leq p \leq \infty$ *and* $0 < s < 1$, *then*

$$(2.34)\qquad \mathcal{S} : B^{p,p}_{-s}(\partial\Omega, \mathcal{F}) \longrightarrow B^{p,p}_{-s+1+1/p}(\Omega, \mathcal{E}), \qquad S : B^{p,p}_{-s}(\partial\Omega, \mathcal{F}) \longrightarrow B^{p,p}_{1-s}(\partial\Omega, \mathcal{E}),$$

are also bounded operators. In fact, the same conclusion applies to

$$(2.35)\qquad \mathcal{S} : B^{p,p}_{-s}(\partial\Omega, \mathcal{F}) \longrightarrow L^p_{-s+1+1/p}(\Omega, \mathcal{E})$$

provided $1 < p < \infty$ *and* $0 < s < 1$. *Also, for* $1 < p < \infty$,

$$(2.36)\qquad \|\mathcal{N}(\mathcal{S}f)\|_{L^p(\partial\Omega)} \leq C(\partial\Omega, p)\|f\|_{L^p_{-1}(\partial\Omega,\mathcal{F})},$$

uniformly for $f \in L^p_{-1}(\partial\Omega, \mathcal{F})$, *whereas for* $1 \leq p \leq \infty$ *and* $0 < s < 1$,

$$(2.37)\qquad \|\mathrm{dist}\,(\cdot, \partial\Omega)^{s-1/p}|\nabla\mathcal{S}f|\,\|_{L^p(\Omega)} \leq C\|f\|_{B^{p,p}_{-s}(\partial\Omega,\mathcal{F})}$$

and

$$(2.38)\qquad \|\mathrm{dist}\,(\cdot, \partial\Omega)^{1/2}|\nabla\mathcal{S}f|\,\|_{L^2(\Omega)} \leq C\|f\|_{L^2_{-1}(\partial\Omega,\mathcal{F})},$$

uniformly in f.

Next, for a first order differential operator $P : \mathcal{E} \to \mathcal{E}$ *with bounded coefficients, we denote by* PS *the principal-value boundary integral operator (in the sense of removing geodesic balls) on* $\partial\Omega$ *with kernel* $P_x E(x,y)$, *then*

$$(2.39)\qquad PS : L^p(\partial\Omega, \mathcal{F}) \longrightarrow L^p(\partial\Omega, \mathcal{E})$$

is well-defined and bounded, and

$$(2.40)\qquad \|\mathcal{N}(P\mathcal{S}f)\|_{L^p(\partial\Omega)} \leq C(\partial\Omega, p)\|f\|_{L^p(\partial\Omega,\mathcal{F})}, \qquad f \in L^p(\partial\Omega, \mathcal{F}), \quad 1 < p < \infty.$$

Moreover,

$$(2.41)\qquad S : \mathfrak{h}^p_{\mathrm{at}}(\partial\Omega, \mathcal{F}) \longrightarrow \mathfrak{H}^{1,p}_{\mathrm{at}}(\partial\Omega, \mathcal{E}), \qquad \tfrac{m-1}{m} < p \leq 1,$$

is well-defined and bounded, and

$$(2.42)\qquad \|\mathcal{N}(\nabla\mathcal{S}f)\|_{L^p(\partial\Omega)} \leq C(\partial\Omega, p)\|f\|_{\mathfrak{h}^p_{\mathrm{at}}(\partial\Omega,\mathcal{F})}, \qquad f \in L^p(\partial\Omega, \mathcal{F}), \quad \tfrac{m-1}{m} < p \leq 1.$$

Also, with $\sigma(L;\xi)$ denoting the principal symbol of L at $\xi \in T^*M$, etc., and with $\Omega_+ := \Omega$, $\Omega_- := M \setminus \bar{\Omega}$

$$(PSf)\big|_{\partial\Omega_\pm} = \mp\tfrac{1}{2}i\sigma(P;\nu)\sigma(L;\nu)^{-1}f + PSf \quad \text{a.e. on } \partial\Omega, \tag{2.43}$$

for each $f \in L^p(\partial\Omega, \mathcal{F})$, $1 < p < \infty$. In fact, similar results are valid at the level of atomic Hardy spaces, provided P has smooth enough coefficients.

Finally, if $\mathcal{E} = \mathcal{F}$ and L is formally self-adjoint, then

$$\begin{aligned} &|\nabla \mathcal{S} f|^2 \operatorname{dist}(\cdot, \partial\Omega)dV \text{ is a Carleson measure} \\ &\qquad \text{on } \Omega \text{ with norm } \leq C\|f\|_{\left(\mathfrak{H}^{1,1}_{\mathrm{at}}(\partial\Omega,\mathcal{E})\right)^*} \end{aligned} \tag{2.44}$$

where dV is the volume element on M.

For the remaining part of the paper, unless otherwise explicitly mentioned, we will always assume that $L : \mathcal{E} \to \mathcal{E}$ is a strongly elliptic, formally self-adjoint operator, whose coefficients satisfy (2.21) and such that the following *non-singularity hypothesis* holds:

$$\forall\, D \subseteq M \text{ Lipschitz},\ u \in L^2_{1,0}(D, \mathcal{E}),\ Lu = 0 \text{ in } D \Longrightarrow u = 0 \text{ in } D. \tag{2.45}$$

It is of relevance to remark that, if L is strongly elliptic, then $L - \lambda$, $\lambda \in \mathbb{R}$, satisfies the non-singularity hypothesis (2.45) provided λ is sufficiently large. This follows from the Gårding inequality, which holds in our setting thanks to (2.21).

We now record an important invertibility result.

PROPOSITION 2.4. *With L as above, for each Lipschitz domain $\Omega \subset M$ there exists $\varepsilon = \varepsilon(\partial\Omega) > 0$ such that*

$$S : L^p_{-s}(\partial\Omega, \mathcal{E}) \longrightarrow L^p_{1-s}(\partial\Omega, \mathcal{E}) \tag{2.46}$$

is invertible for each $p \in (2 - \varepsilon, 2 + \varepsilon)$ and $s \in [0, 1]$.

A proof is contained in [**33**].

In the last part of this section we would like to discuss some representative examples of operators (to which the main results of this paper apply). First, we consider the variable coefficient Lamé operator, which is relevant in the context of the theory of elasticity. To this end, let ∇, Ric be, respectively, the Levi-Civita connection and the Ricci tensor on M, and recall the deformation tensor

$$\operatorname{Def}(X)(Y,Z) := \tfrac{1}{2}\{\langle\nabla_Y X, Z\rangle + \langle\nabla_Z X, Y\rangle\}, \qquad X, Y, Z \in TM. \tag{2.47}$$

Now, for $\mu > 0$ and $\lambda > -2\mu/n$ (where $n := \dim M$), the Lamé operator is given by

$$\mathcal{L}_{\mu,\lambda} := -2\mu \operatorname{Def}^*\operatorname{Def} + \lambda \operatorname{grad}\operatorname{div} = \mu\Delta + (\lambda + \mu)\operatorname{grad}\operatorname{div} + 2\mu \operatorname{Ric}, \tag{2.48}$$

cf., e.g., [**48**]. For Lipschitz domains in the flat Euclidean setting, L^2-results in all space dimensions have been obtained in [**13**], whereas [**12**] contains a sharp atomic theory when $n = 3$. Clearly, $\mathcal{L}_{\mu,\lambda}$ is a second order, strongly elliptic, formally self-adjoint differential operator which, it turns out, satisfies (2.45).

More generally, we can consider operators of the form $L := D^*D$, where $D : \mathcal{E} \to \mathcal{F}$ is an elliptic, first-order differential operator on M, whose (matrix-valued) coefficients are assumed to be sufficiently smooth. If D has the *unique continuation property*, i.e.

$$u \in L^2_1(M),\ Du = 0 \text{ on } M \Rightarrow \operatorname{supp} u \text{ is either } M \text{ or empty}, \tag{2.49}$$

(as is the case for operators of Dirac type) then $L : \mathcal{E} \to \mathcal{E}$ is as in the statement of Proposition 2.2 and, in addition satisfies (2.45).

In closing, let us point out that, whenever self-evident, we shall omit the dependence of various function spaces on the vector bundle $\mathcal{E}$.

3. The single layer potentials on atomic spaces

In this section we retain our standard hypotheses on M, $\mathcal{E}$, Ω and L. In particular, recall that $\dim M = 2$ and $\operatorname{rank}\mathcal{E} = 2$.

PROPOSITION 3.1. *Let ϑ be a $\mathfrak{H}^{1,p}_{at}(\partial\Omega)$-atom and suppose that u is the unique solution of the L^2-Regularity problem $Lu = 0$ in Ω, $\mathcal{N}(\nabla u) \in L^2(\partial\Omega)$, $u|_{\partial\Omega} = \vartheta$. Then there exist a finite constant $\kappa = \kappa(\partial\Omega) > 0$ and some $\varepsilon = \varepsilon(\partial\Omega) \in (0, \frac{1}{6}]$ such that*

$$\int_{\partial\Omega} |\mathcal{N}(\nabla u)|^p \, d\sigma \leq \kappa \tag{3.1}$$

as long as $\frac{2}{3} - \varepsilon < p \leq 1$.

PROOF. We start collecting several useful estimates. First, assume $x_0 \in \partial\Omega$ and $\operatorname{supp}\vartheta \subseteq S_r(x_0) := B_r(x_0) \cap \partial\Omega$, $0 < r < \operatorname{diam}\Omega$. With the dependence of balls on their center occasionally dropped, we have

$$\begin{aligned}
\int_{B_{100r}\cap\partial\Omega} |\mathcal{N}(\nabla u)|^p \, d\sigma &\leq C\Big(\int_{B_{100r}\cap\partial\Omega} |\mathcal{N}(\nabla u)|^2 \, d\sigma\Big)^{p/2} \cdot r^{1-p/2} \\
&\leq C\, r^{1-p/2}\Big(\int_{\partial\Omega} |\mathcal{N}(\nabla u)|^2 \, d\sigma\Big)^{p/2} \\
&\leq C\, r^{1-p/2}\Big(\int_{\partial\Omega} \big[|\partial_\tau\vartheta|^2 + |\vartheta|^2\big] \, d\sigma\Big)^{p/2} \leq \kappa.
\end{aligned} \tag{3.2}$$

The first inequality is Hölder's, the second one is trivial, the third one uses the well-posedness of the L^2-Regularity problem and, finally, the last one is a consequence of the localization and size condition on $\partial_\tau\vartheta$. Next, for $\kappa_o > 0$ as in (2.1) and $0 < R < \operatorname{diam}\Omega$, we introduce two truncated nontangential approach functions

$$\mathcal{N}^R u(x) := \sup\{|u(y)|;\ y \in \Omega,\ R \leq |x - y| \leq \kappa_o \operatorname{dist}(y, \partial\Omega)\}, \tag{3.3}$$

$$\mathcal{N}_R u(x) := \sup\{|u(y)|;\ y \in \Omega,\ |x - y| \leq \kappa_o \operatorname{dist}(y, \partial\Omega),\ |x - y| \leq R\}, \tag{3.4}$$

for $x \in \partial\Omega$, and set $D_R(x) := B_R(x) \cap \Omega$, $S_R(x) := B_R(x) \cap \partial\Omega$, $x \in \partial\Omega$, (again, x is occasionally dropped when irrelevant or obvious). Fix $0 < R < \operatorname{diam}\Omega$ and $x \in \partial\Omega$ such that $S_{4R}(x) \cap S_r(x_0) = \emptyset$. Then, for each $\rho \in [R, 2R]$ write

$$\begin{aligned}\int_{S_\rho}|\mathcal{N}_\rho(\nabla u)|^p\,d\sigma &\leq C\int_{\partial D_{2\rho}}|\mathcal{N}(\nabla u)|^p\,d\sigma\\ &\leq CR^{1-p/2}\Big(\int_{\partial D_{2\rho}}|\mathcal{N}(\nabla u)|^2\,d\sigma\Big)^{p/2}\\ &\leq CR^{1-p/2}\Big(\int_{\Omega\cap\partial B_{2\rho}}|u|^2+|\partial_\tau u|^2\,d\sigma\Big)^{p/2}\\ &\leq CR^{1-p/2}\Big(\int_{\Omega\cap\partial B_{2\rho}}|u|^2+|\nabla u|^2\,d\sigma\Big)^{p/2}.\end{aligned}\tag{3.5}$$

Here $S_\rho = S_\rho(x)$, $D_{2\rho} = D_{2\rho}(x)$, etc., and in the second and third integrals above, $\mathcal{N}$ is taken with respect to the domain in question. The first and the last inequalities are trivial. The second one is based on Hölder, and the third one is obtained using the canonical estimate in the L^2-Regularity problem (cf. [**33**]) plus the fact that $\partial_\tau u = 0$ on $\partial\Omega \cap B_{2\rho}$. Taking the integral average of the extreme sides in (3.5) for $R \leq \rho \leq 2R$ gives

$$\begin{aligned}\int_{S_R}|\mathcal{N}_R(\nabla u)|^p\,d\sigma &\leq CR^{1-p/2}R^{-1}R^{1-p/2}\Big(\iint_{D_{2R}}|u|^2+|\nabla u|^2\Big)^{p/2}\\ &\leq CR^{1-p}R^{-p}\Big(\iint_{D_{4R}}|u|^2\Big)^{p/2} = CR^{1-2p}\Big(\iint_{D_{4R}}|u|^2\Big)^{p/2},\end{aligned}\tag{3.6}$$

by Hölder and boundary Caccioppoli's inequality, respectively; cf. Proposition 2.2.

Next we claim that if $Lu = 0$ in D_{2R}, $u|_{S_{2R}} = 0$ and $\int_{S_{2R}}|\mathcal{N}_R(\nabla u)|^2\,d\sigma < \infty$ then we have the following weak reverse Hölder estimate:

$$\|u\|_{L^\infty(D_R)} \leq C\Big(\frac{1}{R^2}\iint_{D_{2R}}|u|^2\Big)^{1/2}.\tag{3.7}$$

We first prove (3.7) in the special case when $R = 1$. To this end, for $\mu > 0$, we let the subscript μ label the single layer potentials associated with $D_\mu(= B_\mu \cap \Omega)$, and set $g_\mu := S_\mu^{-1}(u|_{\partial D_\mu})$ in $L^2(\partial D_\mu)$. In particular, $u = \mathcal{S}_\mu g_\mu$ in D_μ. Now,

$$\mathcal{S}_\mu : L^2(\partial D_\mu) \longrightarrow L^2_{3/2}(D_\mu) \hookrightarrow L^\infty(D_\mu)\tag{3.8}$$

where the last embedding holds in two dimensions (cf. [**43**]). Hence,

$$\begin{aligned}\|u\|_{L^\infty(D_\mu)} &\leq C\|g_\mu\|_{L^2(\partial D_\mu)} \leq C\|u\|_{L^2_1(\partial D_\mu)}\\ &= C\|u\|_{L^2_1(\partial D_\mu\cap\Omega)} \leq C\|u\|_{L^2(\partial D_\mu\cap\Omega)} + C\|\nabla u\|_{L^2(\partial D_\mu\cap\Omega)}\end{aligned}\tag{3.9}$$

where the equality utilizes $u = 0$ on S_{2R}. Integrating both sides of (3.9) for $1 \leq \mu \leq 3/2$ and invoking boundary Caccioppoli's inequality (i.e. Proposition 2.2) finally yields

$$\|u\|_{L^\infty(D_1)} \leq C\Big(\iint_{D_{3/2}}|u|^2+|\nabla u|^2\Big)^{1/2} \leq C\Big(\iint_{D_2}|u|^2\Big)^{1/2},\tag{3.10}$$

as desired. Turning attention to the general case, i.e. proving (3.7) when $R>0$ is arbitrary, let us observe that if we use the center of S_{2R} as the center of a coordinate system and introduce the dilation operator $v_\rho(x):=v(\rho x)$, then

$$\sum_{j,k}\partial_j A_\rho^{jk}\partial_k u_\rho+\sum_j \rho B_\rho^j\partial_j u_\rho-\rho^2 V_\rho u_\rho=0, \tag{3.11}$$

where

$$A_\rho^{jk}(x):=\big(a_{jk}^{\alpha\beta}(\rho x)\big)_{\alpha,\beta},\ B_\rho^j(x):=\big(b_j^{\alpha\beta}(\rho x)\big)_{\alpha,\beta},\ V_\rho(x):=\big(v^{\alpha\beta}(\rho x)\big)_{\alpha,\beta}. \tag{3.12}$$

Furthermore, the equation (3.11) holds on $D_{2R/\rho}$, for ρ in some interval $(0,\rho_0]$, and $A_\rho^{jk}\xi_j\xi_k\geq\kappa|\xi|^2$ for some $\kappa>0$ uniformly in ρ,ξ, while A_ρ^{jk}, ρB_ρ^j, ρ^2V_ρ are bounded in L_2^r, L_1^r and L^r, respectively. Relying on this, what we have proved so far and the scale-invariant nature of the estimate we seek, (3.7) follows.

Our next claim is that if u satisfies the hypotheses of Proposition 3.1 then

$$\|\mathcal{N}(u)\|_{L^q(\partial\Omega)}\leq C_q r^{1-1/p+1/q},\quad \forall\, q\in(2-\varepsilon,2+\varepsilon). \tag{3.13}$$

Indeed, this is a consequence of the well-posedness of the L^q-Dirichlet problem with q near 2 from [**33**] in concert with the readily verified estimate $\|\vartheta\|_{L^q(\partial\Omega)}\leq Cr^{1-1/p+1/q}$; cf. (2.19).

With (3.2), (3.6), (3.7), and (3.13) taken care of, we are finally ready to tackle (3.1). Let $x\in S_{2R}(x_0)\backslash S_R(x_0)$ and $y\in\Omega$ be such that $R\leq|x-y|\leq\kappa_o\,\mathrm{dist}\,(y,\partial\Omega)$. By interior estimates, cf. [**37**, Proposition 3.4], for each q near 2,

$$\begin{aligned}|\nabla u(y)|&\leq CR^{-1}\Big(\frac{1}{R^2}\iint_{B_{R/2}(y)}|u|^q\Big)^{1/q}+C|u(y)|\\&\leq C\Big(R^{-1}\int_{S_{4R}(x_0)\backslash S_{R/2}(x_0)}|\mathcal{N}(u)|^q\,d\sigma\Big)^{1/q}+C|u(y)|\end{aligned} \tag{3.14}$$

where the last inequality uses Fubini's theorem. Taking supremum over all y's satisfying the aforementioned conditions then gives

$$\mathcal{N}^R(\nabla u)(x)\leq CR^{-1}\Big(\frac{1}{R}\int_{S_{4R}(x_0)\backslash S_{R/2}(x_0)}|\mathcal{N}(u)|^q\,d\sigma\Big)^{1/q}+C\mathcal{N}(u)(x). \tag{3.15}$$

Raise both sides of (3.15) to the p-th power and integrate them (with respect to $d\sigma$) over $x\in S_{2R}(x_0)\setminus S_R(x_0)$ gives

$$\begin{aligned}\int_{S_{2R}(x_0)\backslash S_R(x_0)}|\mathcal{N}^R(\nabla u)|^p\,d\sigma&\leq CR\Big(\frac{1}{R}\int_{S_{4R}(x_0)\backslash S_{R/2}(x_0)}|\mathcal{N}(u)|^q\,d\sigma\Big)^{p/q}\\&\leq CR^{1-p-p/q}\Big(\int_{\partial\Omega}|\mathcal{N}(u)|^q\,d\sigma\Big)^{p/q}.\end{aligned} \tag{3.16}$$

Thanks to (3.13), the last expression above can be bounded by $CR^{1-p-p/q}r^{p-1+p/q}$. All in all,

$$\int_{S_{2R}(x_0)\setminus S_R(x_0)} |\mathcal{N}^R(\nabla u)|^p\,d\sigma \leq C\Big(\frac{r}{R}\Big)^{-1+p+p/q}. \tag{3.17}$$

Next, we seek a similar estimate for $\mathcal{N}_R(\nabla u)$. Assume $R > 100r$. There exist $x_R^{\pm} \in \partial\Omega$ so that $S_{2R}(x_0) \setminus S_R(x_0) \subseteq S_R(x_R^+) \cup S_R(x_R^-)$. Using (3.6), we can write

$$\int_{S_{2R}(x_0)\setminus S_R(x_0)} |\mathcal{N}_R(\nabla u)|^p\,d\sigma \leq CR^{1-2p}\Big(\iint_{D_{3R}(x_R^+)\cup D_{3R}(x_R^-)} |u|^2\Big)^{p/2}. \tag{3.18}$$

Let us now estimate the integral in the right side above assuming that $S_{3R/2}(x_R^{\pm}) \cap S_r(x_0) = \emptyset$ (in the process, we drop the dependence on $x_R^{\pm}$). Inspired by the work in [**4**], introduce

$$\Phi(R) := R^{1/p-2}\Big(\iint_{D_R} |u|^2\Big)^{1/2}. \tag{3.19}$$

Then, with $1 < q < 2 < q' < \infty$ conjugate exponents, Hölder's inequality gives

$$\begin{aligned} \Phi(R)^2 &= R^{2/p-4}\iint_{D_R} |u|^2 \\ &\leq R^{2/p-2}\Big(\frac{1}{R^2}\iint_{D_R} |u|^q\Big)^{1/q}\Big(\frac{1}{R^2}\iint_{D_R} |u|^{q'}\Big)^{1/q'} \\ &=: R^{2/p-2}\, I\cdot II. \end{aligned} \tag{3.20}$$

Now,

$$I \leq C\Big(\frac{1}{R}\int_{S_R} |\mathcal{N}(u)|^q\,d\sigma\Big)^{1/q} \leq C\Big(\frac{1}{R}\int_{\partial\Omega} |\mathcal{N}(u)|^q\,d\sigma\Big)^{1/q} \leq CR^{-1/q}r^{1-1/p+1/q}. \tag{3.21}$$

The inequality (3.21) is a consequence of (3.13) and requires that $2-q$ is small, i.e. $q \in (2-\varepsilon, 2)$ for some small $\varepsilon > 0$ (which we shall assume from now on). As for II, we have thanks to (3.7) and (3.19),

$$II \leq C\|u\|_{L^\infty(D_R)} \leq C\Big(\frac{1}{R^2}\iint_{D_{2R}} |u|^2\Big)^{1/2} \leq CR^{1-1/p}\Phi(2R). \tag{3.22}$$

Putting together (3.20), (3.21) and (3.22) we ultimately obtain

$$\Phi(R)^2 \leq C\Big(\frac{r}{R}\Big)^{\alpha^*}\Phi(2R), \qquad \alpha^* := 1 - 1/p + 1/q. \tag{3.23}$$

Let us also remark that, by virtue of (3.13),

$$\begin{aligned} \Phi(R) &= R^{1/p-2}\Big(\iint_{D_R} |u|^2\Big)^{1/2} \\ &\leq CR^{1/p-2}\Big(R\int_{S_R} |\mathcal{N}u|^2 d\sigma\Big)^{1/2} \leq C\Big(\frac{r}{R}\Big)^{3/2-1/p}. \end{aligned} \tag{3.24}$$

In turn, the estimates (3.23)-(3.24) form the core of a bootstrap argument which we now describe. Let C_0 be a common majorand for the constants appearing in (3.24)-(3.23) and set $\alpha_0 := 3/2 - 1/p$. The iteration scheme we have in mind is based on the observation that, thanks to (3.23),

$$\Phi(R) \leq C_n (r/R)^{\alpha_n} \Longrightarrow \Phi(R) \leq C_{n+1}(r/R)^{\alpha_{n+1}}, \tag{3.25}$$
$$\text{where } C_{n+1} := (C_0 C_n)^{1/2} 2^{-\alpha_n/2}, \ \alpha_{n+1} := (\alpha^* + \alpha_n)/2.$$

Note that $q < 2$ entails $\alpha^* = 1 - 1/p + 1/q > \alpha_0$ so that $\alpha_n \nearrow \alpha^* = 1 - 1/p + 1/q$. Also, it is not difficult to check that $C_n = C_0 2^{-\beta_n}$ where $\beta_n \to \alpha^* = 1 - 1/p + 1/q$. Thus, inductively, $\Phi(R) \leq C_n (r/R)^{\alpha_n}$ holds for each n, uniformly in $r, R > 0$. Passing to the limit $n \to \infty$ then yields

$$\Phi(R) \leq C_0 2^{-1+1/p-1/q} \Big(\frac{r}{R}\Big)^{1-1/p+1/q}. \tag{3.26}$$

Returning with this bound in (3.18) ultimately yields

$$\int_{S_{2R}(x_0)\setminus S_R(x_0)} |\mathcal{N}_R(\nabla u)|^p \, d\sigma \leq C\Big(\frac{r}{R}\Big)^{-1+p+p/q}. \tag{3.27}$$

If we now choose $R := 2^j \cdot 100r$, $j = 0, 1, 2, ...$, we see from (3.17) and (3.27) that

$$\int_{\partial\Omega\setminus S_{100r}} |\mathcal{N}(\nabla u)|^p \, d\sigma \leq C \sum_{j=0}^{\infty} \Big(\frac{1}{2}\Big)^{j(-1+p+p/q)}. \tag{3.28}$$

The above series converges if $-1 + p + p/q > 0$. That is, we need $p > q/(q+1)$ for some $q \in (2-\varepsilon, 2)$. In particular, this is the case whenever $\frac{2}{3} - \varepsilon < p \leq 1$. Together with (3.2), this concludes the proof of (3.1). □

PROPOSITION 3.2. *There exists $\varepsilon = \varepsilon(\partial\Omega) \in (0, \frac{1}{6}]$ such that, for each $\frac{2}{3} - \varepsilon < p \leq 1$, the atomic regularity problem*

$$Lu = 0 \ \text{in } \Omega, \ \ \mathcal{N}(\nabla u) \in L^p(\partial\Omega), \ \ u|_{\partial\Omega} = f \tag{3.29}$$

has a unique solution for each $f \in \mathfrak{H}^{1,p}_{at}(\partial\Omega)$. Moreover, this solution obeys

$$\|\mathcal{N}(\nabla u)\|_{L^p(\partial\Omega)} \leq \kappa \|f\|_{\mathfrak{H}^{1,p}_{at}(\partial\Omega)} \tag{3.30}$$

for some constant $\kappa = \kappa(\partial\Omega) > 0$.

PROOF. Assume that $f = \sum \lambda_j \vartheta_j$ with $\big(\sum |\lambda_j|^p\big)^{1/p} \approx \|f\|_{\mathfrak{H}^{1,p}_{at}(\partial\Omega)}$ and each ϑ_j a $\mathfrak{H}^{1,p}_{at}(\partial\Omega)$-atom. For each j, we let u_j be the solution of the L^2-Regularity problem with boundary datum ϑ_j, as in Proposition 3.1. Then, clearly, $u := \sum \lambda_j u_j$ solves (3.29) and satisfies (3.30). We are therefore left with proving uniqueness, an issue we tackle next.

Suppose that u solves the homogeneous version of (3.29) and let $\Omega_j \nearrow \Omega$ be an approximating sequence as in [**50**]. Then, due to Proposition 2.1 and classical embedding results ([**43**]), $u|_{\partial\Omega_j} \in \mathfrak{H}^{1,p}_{at}(\partial\Omega_j) \hookrightarrow L^{2-\varepsilon}(\partial\Omega_j)$, since $p > \frac{2}{3} - \varepsilon$. Consequently, $\sup_j \int_{\partial\Omega_j} |u|^{2-\varepsilon}\, d\sigma_j < \infty$. If $\varepsilon > 0$ is sufficiently small, it is known that this entails $\mathcal{N}(u) \in L^{2-\varepsilon}(\partial\Omega)$. At this stage, the uniqueness part in the L^p-Dirichlet problem with p near 2 from [**33**] applies and gives $u \equiv 0$ in Ω. This concludes the proof of the proposition. □

We are now ready to tackle the major issue for us here, i.e. the invertibility of the single layer potentials at the atomic level.

THEOREM 3.3. *Let Ω be an arbitrary Lipschitz domain in M. Then there exists $\varepsilon = \varepsilon(\partial\Omega) \in (0, \frac{1}{6}]$ such that the operator*

$$S : \mathfrak{h}^p_{at}(\partial\Omega) \longrightarrow \mathfrak{H}^{1,p}_{at}(\partial\Omega), \quad \textit{for } \tfrac{2}{3} - \varepsilon < p \leq 1, \tag{3.31}$$

is an isomorphism.

PROOF. Recall that $\Omega_+ := \Omega$ and $\Omega_- := M \setminus \bar{\Omega}$. For $f \in \mathfrak{h}^p_{at}(\partial\Omega)$ set $u := \mathcal{S}f$ in $\Omega_\pm$ and recall the conormal derivative from (2.24). Then, with Comp standing for generic compact operators on $\mathfrak{h}^p_{at}(\partial\Omega)$,

$$\begin{aligned}
\|f\|_{\mathfrak{h}^p_{at}(\partial\Omega)} &\leq C\|\partial_\nu u|_{\partial\Omega_+}\|_{\mathfrak{h}^p_{at}(\partial\Omega)} + C\|\partial_\nu u|_{\partial\Omega_-}\|_{\mathfrak{h}^p_{at}(\partial\Omega)} + \|\text{Comp}\,(f)\| \\
&\leq C\|\mathcal{N}(\nabla u|_{\Omega_+})\|_{L^p(\partial\Omega)} + C\|\mathcal{N}(\nabla u|_{\Omega_-})\|_{L^p(\partial\Omega)} + \|\text{Comp}\,(f)\| \\
&\leq C\|u|_{\partial\Omega_+}\|_{\mathfrak{H}^{1,p}_{at}(\partial\Omega)} + C\|u|_{\partial\Omega_-}\|_{\mathfrak{H}^{1,p}_{at}(\partial\Omega)} + \|\text{Comp}\,(f)\| \\
&\leq C\|Sf\|_{\mathfrak{H}^{1,p}_{at}(\partial\Omega)} + \|\text{Comp}\,(f)\|.
\end{aligned} \tag{3.32}$$

The first inequality follows from jumps and the "triangle" inequality, the second one is a consequence of Proposition 2.1, the third one is implied by Proposition 3.2, while the fourth one uses the fact that $u|_{\partial\Omega_-} = u|_{\partial\Omega_+} = Sf$.

Since S is an isomorphism from $L^2(\partial\Omega)$ onto $L^2_1(\partial\Omega)$ (cf. Proposition 2.4), we infer that S in (3.31) is, in fact, onto. Based on this and Lemma 3.4 below, we can then conclude that S is an isomorphism from $\mathfrak{h}^p_{at}(\partial\Omega)$ onto $\mathfrak{H}^{1,p}_{at}(\partial\Omega)$ as long as $\frac{2}{3} - \varepsilon < p \leq 1$. □

Here is the lemma which finishes the proof of Theorem 3.3:

LEMMA 3.4. *Let $\{T_w\}_{w\in U}$ be an interpolating family of operators between two interpolation scales of quasi-Banach spaces, and suppose that U, the space of parameters, is connected and that $\{Y_w\}_{w\in U}$ has the intersection property. Also, assume that T_z is onto for all $z \in U$ and that there exists a point $w^* \in U$ such that $T_{w^*} : X_{w^*} \to Y_{w^*}$ is an isomorphism. Then $T_z : X_z \to Y_z$ is an isomorphism for all $z \in U$.*

This is Theorem 2.10 from [**27**]. The intersection property referred to above formalizes the idea of an interpolation scale whose intersection of all intermediate spaces is fairly rich (cf. [**27**] for all relevant definitions). In the cases which are important for us here (i.e. Banach spaces and atomic Hardy spaces) this is automatically satisfied.

4. The single layer potential on Sobolev-Besov spaces

The goal here is to derive sharp invertibility results for the single layer potential operator on Sobolev-Besov spaces. Once again, we retain the main assumptions from the previous section.

THEOREM 4.1. *For each $\Omega \subset M$ there exists $\varepsilon = \varepsilon(\partial\Omega) > 0$ such that the operators*

$$S : L^p(\partial\Omega) \longrightarrow L^p_1(\partial\Omega), \qquad 1 < p < 2 + \varepsilon, \tag{4.1}$$

and

$$S : L^q_{-1}(\partial\Omega) \longrightarrow L^q(\partial\Omega), \qquad 2 - \varepsilon < q < \infty, \tag{4.2}$$

are isomorphisms.

PROOF. The claim about (4.1) follows by interpolating between the atomic result in Theorem 3.3 and L^p-result with p near 2 from Proposition 2.4. From this, (4.2) also follows by duality. □

To state the next result of this section, for each $0 < \varepsilon \leq \frac{1}{2}$ we let $\mathcal{R}_\varepsilon \subseteq \mathbb{R}^2$ be the region inside the hexagon whose vertices have coordinates $(0,0)$, $(\frac{1}{2}+\varepsilon, 0)$, $(1, \frac{1}{2}-\varepsilon)$, $(1,1)$, $(\frac{1}{2}-\varepsilon, 1)$, $(0, \frac{1}{2}+\varepsilon)$. In particular, $(s, 1/p) \in \mathcal{R}_\varepsilon$ if and only if s, p satisfy either one of the three conditions below:

$$\begin{aligned} &\tfrac{1}{2} - \varepsilon < \tfrac{1}{p} < \tfrac{1}{2} + \varepsilon \ \text{ and } \ 0 < s < 1; \\ &\tfrac{1}{2} + \varepsilon \leq \tfrac{1}{p} < 1 \ \text{ and } \ \tfrac{1}{p} - \tfrac{1}{2} - \varepsilon < s < 1; \\ &0 < \tfrac{1}{p} \leq \tfrac{1}{2} - \varepsilon \ \text{ and } \ 0 < s < \tfrac{1}{p} + \tfrac{1}{2} + \varepsilon. \end{aligned} \tag{4.3}$$

THEOREM 4.2. *Let $\Omega \subset M$ be an arbitrary Lipschitz domain. Then there exists $\varepsilon(\partial\Omega) > 0$ such that the operator*

$$S : B^{q,q}_{-s}(\partial\Omega) \longrightarrow B^{q,q}_{1-s}(\partial\Omega), \tag{4.4}$$

is an isomorphism for each $(s, 1/p) \in \mathcal{R}_\varepsilon$, $1/p + 1/q = 1$.

PROOF. We shall need a few results from functional analysis which we briefly recall. If X is a quasi-Banach space with a separating dual, then the completion of X in the Mackey topology (i.e. the strongest locally convex topology on X which produces the same topological dual as the original topology on X), is denoted by $\hat{X}$ and is called the Banach envelope of X. It is not difficult to show that, for a linear and bounded operator T on X, the property of being an isomorphism is preserved by applying "hat".

Quite recently, it has been proved in **[31]** that if $\Omega \subset M$ is a two-dimensional Lipschitz domain then

$$\widehat{\mathfrak{h}^p_{at}(\partial\Omega)} = B^{1,1}_{1-1/p}(\partial\Omega), \qquad \widehat{\mathfrak{H}^{1,p}_{at}(\partial\Omega)} = B^{1,1}_{2-1/p}(\partial\Omega), \qquad \forall p \in (\tfrac{1}{2}, 1). \tag{4.5}$$

The strategy for proving Theorem 4.2 is to apply "hat" to the results in Theorem 3.3 and then to interpolate. More specifically, (3.31), (4.5) and duality give that the following are isomorphism for each $\frac{2}{3} - \varepsilon < p < 1$:

$$(4.6) \quad S : B^{1,1}_{1-1/p}(\partial\Omega) \longrightarrow B^{1,1}_{2-1/p}(\partial\Omega), \qquad S : \left(B^{1,1}_{2-1/p}(\partial\Omega)\right)^* \longrightarrow B^{\infty,\infty}_{1/p-1}(\partial\Omega).$$

Then interpolation between (4.6) and the previously known L^2-results (cf. Proposition 2.4) yields precisely the range $\mathcal{R}_\varepsilon$ specified in the statement of the theorem. □

5. Applications

Once more we retain our standard hypotheses on M, $\mathcal{E}$, Ω and L; in particular, $\dim M = 2$ and $\operatorname{rank}\mathcal{E} = 2$. First we consider the classical Poisson problems for L with Dirichlet boundary conditions on arbitrary two-dimensional Lipschitz domains.

THEOREM 5.1. *Let Ω be a (two dimensional) Lipschitz domain in M. Then there exists $\varepsilon = \varepsilon(\partial\Omega) > 0$ such that if $1 < p, q < \infty$, $1/p + 1/q = 1$, $0 < s < 1$, are such that $(s, 1/p) \in \mathcal{R}_\varepsilon$, the Poisson problem for the L with Dirichlet boundary condition:*

$$(5.1) \qquad \begin{cases} u \in L^p_{s+\frac{1}{p}}(\Omega, \mathcal{E}), \\ Lu = f \in L^p_{\frac{1}{p}+s-2}(\Omega, \mathcal{E}), \\ \operatorname{Tr} u = g \in B^{p,p}_s(\partial\Omega, \mathcal{E}), \end{cases}$$

is well-posed for each $(s, 1/p) \in \mathcal{R}_\varepsilon$. In general, if L does not necessarily satisfy the nonsingularity hypothesis (2.45), then (5.1) is Fredholm with index zero.

PROOF. Assume first that L satisfies the nonsingularity hypothesis (2.45). Then, existence is sought in the form $u := L^{-1}(\tilde{f}) + \mathcal{S}h$ where $\tilde{f} \in L^p_{\frac{1}{p}+s-2}(M)$ is a suitable extension of f and $h := S^{-1}(g - \operatorname{Tr} L^{-1}(\tilde{f})) \in B^{p,p}_{s-1}(\partial\Omega)$. Note that this also yields natural estimates. Thus, so far,

$$(5.2) \qquad L_D : L^p_{s+\frac{1}{p}}(\Omega) \longrightarrow B^{p,p}_s(\partial\Omega) \times \left(L^{p'}_{s'+\frac{1}{p'},0}(\Omega)\right)^*, \quad L_D u := (\operatorname{Tr} u, Lu),$$

where $s' := 1 - s$, $1/p + 1/p' = 1$, is well-defined, linear, bounded and onto for each $(s, 1/p) \in \mathcal{R}_\varepsilon$. Since, by virtue of (2.45), this operator is also one-to-one when $p = 2$ and $s = \frac{1}{2}$, Lemma 3.4 gives that L_D in (5.2) is actually an isomorphism for each $(s, 1/p) \in \mathcal{R}_\varepsilon$.

When L does not necessarily satisfy (2.45), pick $\lambda > 0$ large enough so that $L - \lambda$ fulfills the nonsingularity hypothesis (2.45). Then the operator L_D in (5.2) is seen to be a compact perturbation of $(L - \lambda)_D$ which is invertible and, hence, L_D is Fredholm with index zero. □

Next we consider the L^p-Dirichlet and Regularity problems for L in Dahlberg's sense.

THEOREM 5.2. *Let Ω be a Lipschitz domain in M. Then there exists $\varepsilon = \varepsilon(\partial\Omega) > 0$ such that if $1 < p < 2 + \varepsilon$, the regularity problem*

$$(5.3)\qquad \begin{cases} u \in C^1_{\rm loc}(\Omega, \mathcal{E}), \\ Lu = 0 \ in \ \Omega, \\ \mathcal{N}(\nabla u) \in L^p(\partial\Omega), \\ \mathrm{Tr}\ u = f \in L^p_1(\partial\Omega, \mathcal{E}), \end{cases}$$

and, for $2 - \varepsilon < q < \infty$, *the Dirichlet problem*

$$(5.4)\qquad \begin{cases} u \in C^1_{\rm loc}(\Omega, \mathcal{E}), \\ Lu = 0 \ in \ \Omega, \\ \mathcal{N}(u) \in L^q(\partial\Omega), \\ \mathrm{Tr}\ u = g \in L^q(\partial\Omega, \mathcal{E}), \end{cases}$$

are well-posed.

PROOF. In the case of (5.3), existence plus estimates follow from Theorem 4.1, by taking $u := \mathcal{S}(S^{-1}f)$. Uniqueness is then seen much as in [**33**, §3]. A similar reasoning applies to (5.4), this time relying on (4.2). □

Our next application deals with the case when

$$(5.5)\qquad -\langle Lu, u\rangle \geq \kappa \|\nabla u\|^2_{L^2(\Omega)}, \qquad \forall\, u \in L^2_{1,0}(\Omega),$$

for some $\kappa > 0$, independent of u. We shall be concerned with $A := \sqrt{-L}$, the functional analytic square-root of the solution operator for the problem (5.1) with *homogeneous* boundary conditions.

THEOREM 5.3. *Let* Ω *be a Lipschitz domain in* M, $\dim M = 2$, *and* L *as above (including (5.5)). Then there exists* $\varepsilon = \varepsilon(\Omega, L) > 0$ *such that the operator*

$$(5.6)\qquad A^{-r} : L^q(\Omega) \longrightarrow L^q_{r,0}(\Omega)$$

is an isomorphism for every r, q *satisfying*

$$(5.7)\qquad 0 < r < 2, \quad 1 < q < \infty, \quad \max\{r/2 - 1/4 - \varepsilon, r - 1\} \leq 1/q.$$

PROOF. As in [**25**], [**30**], where the case of the square-root of the Euclidean Dirichlet Laplacian is discussed, our proof relies on three basic isomorphisms, i.e.

$$(5.8)\qquad A : L^2_{1,0}(\Omega) \xrightarrow{\sim} L^2(\Omega),$$

$$(5.9)\qquad A^{i\delta} : L^p(\Omega) \xrightarrow{\sim} L^p(\Omega), \quad 1 < p < \infty,\ \delta \in \mathbb{R},$$

$$(5.10)\qquad A^{-2} : L^p_{s+1/p-2}(\Omega) \xrightarrow{\sim} L^p_{s+1/p,0}(\Omega), \quad (s, 1/p) \in \mathcal{R}_\varepsilon.$$

The first one follows much as in the case of the Laplacian, the second one is a consequence of Stein's Littlewood-Paley multiplier theory for semi-groups in [**46**] (cf. also Corollary 1 in [**8**]), while the third one is implied by Theorem 5.1.

For starters, by interpolating between the cases $r = 0$, $1 < q < \infty$ and $r = 1$, $q = 2$ (with the aid of Stein's interpolation theorem for analytic families of operators; [**47**]), we arrive at the conclusion that

$$(5.11) \qquad (5.6) \text{ is an isomorphism for } (r, 1/q) \in \mathcal{T},$$

where $\mathcal{T}$ is the (open) triangle with vertices at $(0,0)$, $(1,1/2)$ and $(0,1)$.

The strategy is to successively enhance this region of validity for (5.6) via a procedure based on interpolation and an observation which we now describe. Introduce $\Phi(x,y) := (2-x, 1-y)$, i.e. the symmetry with respect to $(1,1/2)$, as well as $\Psi(x,y) := (2-x-y, 1-y)$. Then, so we claim, for each set $\mathcal{O} \subset \Psi(\mathcal{R}_\varepsilon)$,

$$(5.12) \qquad \begin{array}{l} (5.6) \text{ an isomorphism } \forall\, (r,1/q) \in \mathcal{O} \Longrightarrow \\ (5.6) \text{ is also an isomorphism } \forall\, (r,1/q) \in \Phi(\mathcal{O}). \end{array}$$

To see this, assume that (5.6) is an isomorphism for some $(r,1/q) \in \Psi(\mathcal{R}_\varepsilon)$. Composing this with (5.9) yields that $A^{-r+i\delta} : L^q(\Omega) \xrightarrow{\sim} L^q_{r,0}(\Omega)$ is an isomorphism and, further, dualizing its inverse, that $A^{r+i\delta} : L^{q'}(\Omega) \xrightarrow{\sim} L^{q'}_{-r}(\Omega)$, $1/q + 1/q' = 1$, is an isomorphism, for each $\delta \in \mathbb{R}$. At this point, we would like to compose this last isomorphism with (5.10), in order to obtain that

$$(5.13) \qquad A^{-\mu+i\delta} : L^{q'}(\Omega) \xrightarrow{\sim} L^{q'}_{\mu,0}(\Omega), \quad \mu := 2-r, \quad \delta \in \mathbb{R}$$

is an isomorphism. For this to work we need $q' = p$ and $-r = s + 1/p - 2$ for some $(s, 1/p) \in \mathcal{R}_\varepsilon$, i.e. $(2 - r - 1/q', 1/q') \in \mathcal{R}_\varepsilon$, or $\Psi^{-1}(r, 1/q) \in \mathcal{R}_\varepsilon$. Note that this is automatically taken care of by the fact that $(r,1/q) \in \mathcal{O} \subset \Psi(\mathcal{R}_\varepsilon)$. Observing now that $(\mu, 1/q') = \Phi(r, 1/q)$ concludes the proof of (5.12).

With (5.11) and (5.12) taken care of, it is now straightforward to finish the proof of the theorem. The remaining steps are as follows. Call a subset D of $(0,2) \times (0,1)$ 'good' if (5.6) is an isomorphism for each $(r,1/q) \in D$. Thus, in this terminology, $\mathcal{T}$ from (5.11) is good. Next, use (5.12) with $\mathcal{O} := \mathcal{T} \cap \Psi(\mathcal{R}_\varepsilon)$ so that $\mathcal{T}_1 := \Phi(\mathcal{T} \cap \Psi(\mathcal{R}_\varepsilon)) = \Phi(\mathcal{T}) \cap \Psi(\mathcal{R}_\varepsilon)$ is also good. By interpolation, the convex hull of $\mathcal{T}$ and $\mathcal{T}_1$, call it $\mathcal{T}_2$, is also good. Relying again on (5.12), this time with $\mathcal{O} := \mathcal{T}_2 \cap \Psi(\mathcal{R}_\varepsilon)$, proves that $\mathcal{T}_3 := \Phi(\mathcal{T}_2 \cap \Psi(\mathcal{R}_\varepsilon)) = \Phi(\mathcal{T}_2) \cap \Psi(\mathcal{R}_\varepsilon)$ is good. Finally, interpolating between $\mathcal{T}_2$ and $\mathcal{T}_3$ yields precisely the region (5.7). □

The important case $r = 1$, allowing for a direct comparison of $\sqrt{-L}$ with the ordinary gradient, deserves to be stated separately.

THEOREM 5.4. *Let Ω be a Lipschitz domain in M, $\dim M = 2$, and $L \le 0$ as above. There*

$$(5.14) \qquad \sqrt{-L} : L^p_{1,0}(\Omega) \xrightarrow{\sim} L^p(\Omega)$$

is well-defined and bounded for each $1 < p < \infty$, and an isomorphism for each $1 < p < 4+\varepsilon$, where $\varepsilon = \varepsilon(\partial\Omega, L) > 0$. This result is sharp in the class of Lipschitz domains.

Similar results have been obtained for the Dirichlet Laplacian in Theorem 7.5 of [**25**] and for the Neumann Laplacian in [**31**]. These results have been further extended in [**37**].

The above theorem can further be used to obtain mapping properties for the Green function of L on L^p-spaces.

THEOREM 5.5. *Once again assume that L satisfies (5.5) and consider $G(x,y)$, the Green function for L in the Lipschitz domain Ω. Finally, denote by G the operator sending f to $\iint_\Omega G(\cdot,y)f(y)\,dV_y$. Then, for some $\varepsilon=\varepsilon(\Omega)>0$,*

$$\left(\iint_\Omega |\nabla Gf|^q\right)^{\frac{1}{q}} \le C\left(\iint_\Omega |f|^p\right)^{\frac{1}{p}}, \quad f\in L^p(\Omega), \tag{5.15}$$

provided $1<p<q<4+\varepsilon$, $\frac{1}{q}=\frac{1}{p}-\frac{1}{2}$. This result is sharp in the class of Lipschitz domains.

This is the two dimensional, system counterpart of a well-known result of Dahlberg ([**10**]) concerning estimates for harmonic Green potentials in Lipschitz domains. As in [**25**] (cf. also [**30**], [**37**]) this follows by using Theorem 5.14, Sobolev's embedding theorem and the factorization

$$G : L^p(\Omega) \xrightarrow{A^{-1}} L^p_{1,0}(\Omega) \overset{\iota}{\hookrightarrow} L^q(\Omega) \xrightarrow{A^{-1}} L^q_{1,0}(\Omega). \tag{5.16}$$

Before going any further, we would like to point out that a weak maximum principle, i.e. the estimate $\|u\|_{L^\infty(\Omega)} \le C\|u\|_{L^\infty(\partial\Omega)}$ for null-solutions of L, can be proved as in [**12**] based on what is available so far. In turn, the solvability of the Dirichlet problem for L with continuous boundary data also follows. The same circle of ideas further yield the fact that the single layer

$$\mathcal{S} : L^p_{-s}(\partial\Omega,\mathcal{E}) \longrightarrow L^p_{1-s+1/p}(\partial\Omega,\mathcal{E}) \tag{5.17}$$

is a well-defined and bounded operator for each $2\le p<\infty$; cf. [**37**, Theorem 8.7].

To state our next result, recall that the Laplace-Beltrami operator Δ on M is given in local coordinates by

$$\Delta u := (\det(g_{jk}))^{-1/2}\sum_j \partial_j\Big(\sum_k g^{jk}(\det(g_{jk}))^{1/2}\,\partial_k u\Big), \tag{5.18}$$

where we take (g^{jk}) to be the matrix inverse of (g_{jk}). Also, for $V\ge 0$, $V\in L^\infty(M)$, we denote by $E(x,y)$ the Schwartz kernel of $(\Delta - V)^{-1}$ and let

$$Kf(x) := \text{p.v.}\int_{\partial\Omega} \frac{\partial E}{\partial\nu_y}(x,y)\,f(y)\,d\sigma(y), \qquad x\in\partial\Omega. \tag{5.19}$$

Here p.v. indicates that the integral is taken in the principal value sense. Specifically, we fix a smooth background metric which, in turn, induces a distance function on M. In particular, we can talk about balls and p.v. is defined in the usual sense, of removing such small geodesic balls and passing to the limit. Moreover, we denote by K^* the formal transpose of K.

That K, K^* are singular integral operators of Calderón-Zygmund type follows, as in [**35**], from the estimates on $E(x,y)$ contained below.

PROPOSITION 5.6. *Assume that the metric tensor on M has components of class C^α for some $\alpha > 0$. Then, in local coordinates in which the metric tensor is given by $g = \sum_{j,k} g_{jk} dx_j \otimes dx_k$, we have*

$$\begin{aligned} E(x,y) &= \frac{1}{\sqrt{\det(g_{jk}(y))}}\Big\{e_0(x-y,y)+e_1(x,y)\Big\} \\ &= \frac{1}{\sqrt{\det(g_{jk}(x))}}\Big\{e_0(x-y,x)+e_1(y,x)\Big\}, \end{aligned} \tag{5.20}$$

where the principal part is given by

$$e_0(x-y,y) = -\frac{1}{2\pi}\log\Big(\sum g_{jk}(y)(x_j-y_j)(x_k-y_k)\Big)^{1/2}, \tag{5.21}$$

while the remainder $e_1(x,y)$ is bounded and satisfies

$$|\nabla_x e_1(x,y)| \leq C_\varepsilon |x-y|^{-1+\alpha-\varepsilon} \tag{5.22}$$

for each $\varepsilon > 0$ small.

PROOF. The argument closely parallels that in [**36**], [**37**], [**38**] where the case $\dim M \geq 3$ is discussed. □

Our last theorem in this section deals with invertibility results for the harmonic layer potential operators on Sobolev-Besov spaces. Recall that $\mathcal{C}$ stands for the collection of all constant functions on $\partial\Omega$ and set

$$\tilde{B}^{p,p}_{-s}(\partial\Omega) := \{f \in B^{p,p}_{-s}(\partial\Omega) : \langle f, \chi\rangle = 0, \ \ \forall \chi \in \mathcal{C}\}, \tag{5.23}$$

for $1 \leq p \leq \infty$, $0 < s < 1$.

THEOREM 5.7. *Suppose that M is a Riemannian manifold equipped with a metric tensor of class $C^{1+\gamma}$, $\gamma > 0$ (for specific results, this assumption can be further relaxed; cf. [**38**]) and let Ω be an arbitrary, connected Lipschitz domain in M. Then, there exists $\varepsilon = \varepsilon(\partial\Omega) \in (0, \frac{1}{2}]$ such that the operators*

(1) $\pm\frac{1}{2}I + K : L^p(\partial\Omega) \longrightarrow L^p(\partial\Omega), \quad 2-\varepsilon < p < \infty;$
(2) $\pm\frac{1}{2}I + K : B^{p,p}_s(\partial\Omega) \longrightarrow B^{p,p}_s(\partial\Omega), \quad (s, 1/p) \in \mathcal{R}_\varepsilon;$
(3) $\pm\frac{1}{2}I + K^* : L^p(\partial\Omega) \longrightarrow L^p(\partial\Omega), \quad 1 < p < 2+\varepsilon;$
(4) $\pm\frac{1}{2}I + K^* : B^{q,q}_{-s}(\partial\Omega) \longrightarrow B^{q,q}_{-s}(\partial\Omega), \quad (s, 1/p) \in \mathcal{R}_\varepsilon, \ \frac{1}{q} = 1 - \frac{1}{p},$

are Fredholm, of index zero. Furthermore, assuming that $V > 0$ on a set of positive measure in each connected component of $M \setminus \overline{\Omega}$, the operators

(5) $\frac{1}{2}I + K : L^p(\partial\Omega) \longrightarrow L^p(\partial\Omega), \quad 2-\varepsilon < p < \infty;$
(6) $\frac{1}{2}I + K : B^{p,p}_s(\partial\Omega) \longrightarrow B^{p,p}_s(\partial\Omega), \quad (s, 1/p) \in \mathcal{R}_\varepsilon;$
(7) $\frac{1}{2}I + K^* : L^p(\partial\Omega) \longrightarrow L^p(\partial\Omega), \quad 1 < p < 2+\varepsilon;$
(8) $\frac{1}{2}I + K^* : B^{q,q}_{-s}(\partial\Omega) \longrightarrow B^{q,q}_{-s}(\partial\Omega), \quad (s, 1/p) \in \mathcal{R}_\varepsilon, \ \frac{1}{q} = 1 - \frac{1}{p};$
(9) $\frac{1}{2}I + K : L^p_1(\partial\Omega) \longrightarrow L^p_1(\partial\Omega), \quad 1 < p < 2+\varepsilon;$
(10) $\frac{1}{2}I + K^* : L^p_{-1}(\partial\Omega) \longrightarrow L^p_{-1}(\partial\Omega), \quad 2-\varepsilon < p < \infty,$

are invertible. Also, if $V > 0$ on a set of positive measure in Ω, then

(11) $-\frac{1}{2}I + K : L^p(\partial\Omega) \longrightarrow L^p(\partial\Omega), \quad 2-\varepsilon < p < \infty;$

(12) $-\frac{1}{2}I + K : B^{p,p}_s(\partial\Omega) \longrightarrow B^{p,p}_s(\partial\Omega)$, $(s, 1/p) \in \mathcal{R}_\varepsilon$;
(13) $-\frac{1}{2}I + K^* : L^p(\partial\Omega) \longrightarrow L^p(\partial\Omega)$, $1 < p < 2 + \varepsilon$;
(14) $-\frac{1}{2}I + K^* : B^{q,q}_{-s}(\partial\Omega) \longrightarrow B^{q,q}_{-s}(\partial\Omega)$, $(s, 1/p) \in \mathcal{R}_\varepsilon$, $\frac{1}{q} = 1 - \frac{1}{p}$;
(15) $-\frac{1}{2}I + K : L^p_1(\partial\Omega) \longrightarrow L^p_1(\partial\Omega)$, $1 < p < 2 + \varepsilon$,

are isomorphisms, while, if $V = 0$ *on* $\bar{\Omega}$, *then*

(16) $-\frac{1}{2}I + K : L^p(\partial\Omega)/\mathcal{C} \longrightarrow L^p(\partial\Omega)/\mathcal{C}$, $2 - \varepsilon < p < \infty$;
(17) $\pm\frac{1}{2}I + K : B^{p,p}_s(\partial\Omega)/\mathcal{C} \longrightarrow B^{p,p}_s(\partial\Omega)/\mathcal{C}$, $(s, 1/p) \in \mathcal{R}_\varepsilon$;
(18) $-\frac{1}{2}I + K^* : L^p_0(\partial\Omega) \longrightarrow L^p_0(\partial\Omega)$, $1 < p < 2 + \varepsilon$;
(19) $\pm\frac{1}{2}I + K^* : \tilde{B}^{q,q}_{-s}(\partial\Omega) \longrightarrow \tilde{B}^{q,q}_{-s}(\partial\Omega)$, $(s, 1/p) \in \mathcal{R}_\varepsilon$, $\frac{1}{q} = 1 - \frac{1}{p}$;
(20) $-\frac{1}{2}I + K : L^p_1(\partial\Omega)/\mathcal{C} \longrightarrow L^p_1(\partial\Omega)/\mathcal{C}$, $1 < p < 2 + \varepsilon$;
(21) $S : \tilde{B}^{q,q}_{-s}(\partial\Omega) \longrightarrow B^{q,q}_{1-s}(\partial\Omega)/\mathcal{C}$, $(s, 1/p) \in \mathcal{R}_\varepsilon$, $\frac{1}{q} = 1 - \frac{1}{p}$,

are isomorphisms, where $\mathcal{C}$ *is the space of constant functions on* $\partial\Omega$ *and* $L^p_0(\partial\Omega)$ *consists of elements of* $L^p(\partial\Omega)$ *integrating to zero.*

In the class of Lipschitz domains, these results are sharp. If, however, $\partial\Omega \in C^1$, *then we can take* $1 < p < \infty$, $1/p + 1/q = 1$ *and* $0 < s < 1$.

PROOF. The departure point (seen from §4 much as in [**32**]) is the observation that

$$\pm\tfrac{1}{2}I + K^* : \mathfrak{h}^p_{\rm at}(\partial\Omega) \to \mathfrak{h}^p_{\rm at}(\partial\Omega), \qquad \pm\tfrac{1}{2}I + K : \mathfrak{H}^{1,p}_{\rm at}(\partial\Omega) \to \mathfrak{H}^{1,p}_{\rm at}(\partial\Omega) \tag{5.24}$$

are Fredholm operators with index zero for each $\frac{2}{3} - \varepsilon < p \le 1$. Taking Banach envelopes and dualizing, this gives that

$$\pm\tfrac{1}{2}I + K : B^{\infty,\infty}_s(\partial\Omega) \to B^{\infty,\infty}_s(\partial\Omega), \qquad \pm\tfrac{1}{2}I + K : B^{1,1}_{1-s}(\partial\Omega) \to B^{1,1}_{1-s}(\partial\Omega) \tag{5.25}$$

are Fredholm operators with index zero for each $0 < s < \frac{1}{2} + \varepsilon$. With these at hand, repeated applications of real and complex interpolation yield the desired conclusion. □

Given these results, solutions to the Dirichlet, Regularity and Neumann problems for the operator Δ in Lipschitz subdomains of M can be produced for L^p and Besov boundary data, for appropriate (sharp ranges of) indices and optimal estimates. We omit the details.

6. The three dimensional case

While up to this point we have focused almost exclusively on the case $\dim M = 2$, here we would like to comment on the case $\dim M = 3$. The idea is that our methods are flexible enough to yield optimal results in this setting, and most proofs parallel closely those developed when $\dim M = 2$. The general principle is that all major results continue to hold in this latter setting, albeit for possibly smaller (yet sharp) ranges of indices. Concretely, the three-dimensional version of Theorem 5.1 requires that $0 < s < 1$ and $1 < p < \infty$ satisfy at least one of the conditions:

$$
(6.1) \qquad \begin{aligned}
&\tfrac{2}{1+\varepsilon} < p < \tfrac{2}{1-\varepsilon} \quad \text{and} \quad 0 < s < 1; \\
&1 \leq p < \tfrac{2}{1+\varepsilon} \quad \text{and} \quad \tfrac{2}{p} - 1 - \varepsilon < s < 1; \\
&\tfrac{2}{1-\varepsilon} < p \leq \infty \quad \text{and} \quad 0 < s < \tfrac{2}{p} + \varepsilon,
\end{aligned}
$$

while Theorem 5.2 goes through unchanged. In three dimensions, the last double inequality in (5.7) of Theorem 5.4 should be changed to $\max\{r/3 - \varepsilon, r - 1\} \leq 1/q$. Next, the three-dimensional version of Theorem 5.4 only requires changing the range for which (5.14) is an isomorphism to $1 < p < 3 + \varepsilon$. In the three-dimensional context, the validity range for Theorem 5.5 is $1 < p < q < 3 + \varepsilon$, $1/q = 1/p - 1/3$, while in Theorem 5.7 the region $\mathcal{R}_\varepsilon$ should refer to (6.1). In this form, all the aforementioned results continue to be sharp in the class of three-dimensional Lipschitz domains.

In closing, we would like to point out that by relying on the results proved in §4-§6, other significant problems can be successfully attacked. As examples, we list the issue of deriving estimates for complex powers of systems of PDE's (as in [**30**]), establishing sharp Hodge decompositions (as in [**32**]), and producing existence results for semilinear versions of (5.1), as in [**16**].

References

[1] C. Bennett and R. Sharpley, *Interpolation of Operators*, Academic Press, Boston, 1988.

[2] J. Bergh and J. Löfström, *Interpolation Spaces. An Introduction*, Springer-Verlag, Berlin/New York, 1976.

[3] R. Brown, *The initial-Neumann problem for the heat equation in Lipschitz cylinders*, Trans. Amer. Math. Soc., Vol. 320, No. 1 (1990), 1–52.

[4] R. Brown and Z. Shen, *Boundary value problems in Lipschitz cylinders for three-dimensional parabolic systems*, Revista Matematica Iberoamericana, Vol. 8, No. 3 (1992), 271–303.

[5] A. P. Calderón, *Boundary value problems for the Laplace equation in Lipschitz domains*, Recent progress in Fourier Analysis, pp. 33–48, I. Peral and J.-L. Rubio de Francia eds., Elsevier Science Pub., North-Holland, 1985.

[6] Y.-Z. Chen and L.-C. Wu, *Second Order Elliptic Equations and Elliptic Systems*, AMS Mathematical Monographs, Vol. 174, Providence, RI, 1998.

[7] R. Coifman and G. Weiss, *Extensions of Hardy spaces and their use in analysis*, Bull. Amer. Math. Soc. 83 (1977), 569–645.

[8] M. G. Cowling, *Harmonic analysis on semigroups*, Annals of Mathematics, Vol. 117 (1983), 267–283.

[9] B. Dahlberg, *On estimates of harmonic measure*, Arch. Rat. Mech. Anal. 65 (1977), 275–288.

[10] B. E. Dahlberg, *L^q-estimates for Green potentials in Lipschitz domains*, Mathematica Scandinavica, 44 (1979), 149–170.

[11] B. Dahlberg and C. Kenig, *Hardy spaces and the L^p-Neumann problem for Laplace's equation in a Lipschitz domain*, Ann. Math. 125 (1987), 437–465.

[12] B. Dahlberg and C. Kenig, *L^p Estimates for The Three-dimensional System of Ellastostatics on Lipschitz Domains*, Lecture Notes Pure Appl. Math., Vol. 122, 1990, 621–634.

[13] B. Dahlberg, C. Kenig and G. Verchota, *Boundary value problems for the system of elastostatics on Lipschitz domains*, Duke Math. J., 57 (1988), 795–818.

[14] S. Dahlke and R. A. DeVore, *Besov regularity for elliptic boundary value problems*, Comm. PDE, 22 (1997), 1–16.

[15] S. Dahlke, *Besov regularity for second order elliptic boundary value problems with variable coefficients*, Manuscripta Math., 95 (1998), 59–77.

[16] M. Dindoš and M. Mitrea, *Semilinear Poisson problems in Sobolev-Besov spaces on Lipschitz domains*, preprint (2000).

[17] E. Fabes, *Layer potential methods for boundary value problems on Lipschitz domains*, Potential Theory, Surveys and Problems, J. Král et al eds., Springer-Verlag Lecture Notes in Math., No. 1344, 1988, 55–80.

[18] E. Fabes, C. Kenig, and G. Verchota, *The Dirichlet problem for the Stokes system on Lipschitz domains*, Duke Math. J., 57 (1988), 769–793.

[19] E. Fabes, M. Jodeit and N. Rivière, *Potential techniques for boundary value problems on C^1 domains*, Acta Math. 141 (1978), 165–186.

[20] E. Fabes, O. Mendez and M. Mitrea, *Boundary layers on Sobolev-Besov spaces and Poisson's equation for the Laplacian in Lipschitz domains*, J. Funct. Anal. 159 (1998), 323–368.

[21] U. Frank, *A recurring theorem about pairs of quadratic forms and extensions: a survey*, Linear Algebra Appl., 25 (1979), 219–237.

[22] M. Frazier and B. Jawerth, *Decomposition of Besov spaces*, Indiana Univ. Math. J., 34 (1985), 777–799.

[23] S. Hofmann and J. Lewis, *L^2-Solvability and representation by caloric layer potentials in time-varying domains*, Annals of Math., 144 (1996), 349–420.

[24] D. Jerison and C. E. Kenig, *The Neumann problem on Lipschitz domains*, Bull. Amer. Math. Soc. 4 (1981), 203–207.

[25] D. Jerison and C. E. Kenig, *The Inhomogeneous Dirichlet Problem in Lipschitz Domains*, J. Funct. Anal., Vol. 127, No. 1 (1995), 204–234.

[26] A. Jonsson and H. Wallin, *Function Spaces on Subsets of $\mathbb{R}^n$*, Harwood Academic, New York, 1984.

[27] N. Kalton and M. Mitrea, *Stability results on interpolation scales of quasi-Banach spaces and applications*, Trans. Amer. Math. Soc., Vol. 350, No. 10 (1998), 3903–3922.

[28] C. E. Kenig, *Elliptic boundary value problems on Lipschitz domains*, Beijing lectures in harmonic analysis (Beijing, 1984), 131–183, Ann. of Math. Stud., 12, Princeton Univ. Press, Princeton N.J., 1986.

[29] C. Kenig, *Harmonic Analysis Techniques for Second Order Elliptic Boundary Value Problems*, AMS Regional Conferences Series in Mathematics, Vol. 83, 1994.

[30] O. Mendez and M. Mitrea, *Complex powers of the Neumann Laplacian in Lipschitz domains*, to appear in Mathematische Nachrichten, (2001).

[31] O. Mendez and M. Mitrea, *The Banach envelope of Besov and Triebel-Lizorkin spaces and applications*, J. Fourier Anal. Appl. Vol. 6 No. 5 (2000), 503–531.

[32] D. Mitrea, *Layer potentials and Hodge decompositions in two dimensional Lipschitz domains*, to appear in Math. Annalen, (2001).

[33] D. Mitrea, M. Mitrea and M. Taylor, *Layer potentials, the Hodge Laplacian, and global boundary problems in nonsmooth Riemannian manifolds*, to appear in Memoirs of AMS, 2001.

[34] M. Mitrea, *The method of layer potentials in electromagnetic scattering theory on nonsmooth domains*, Duke Math. J., 77 (1995), 111–133.

[35] M. Mitrea and M. Taylor, *Boundary layer methods for Lipschitz domains in Riemannian manifolds*, J. Funct. Anal., Vol. 163 (1999), 181–251.

[36] M. Mitrea and M. Taylor, *Potential theory on Lipschitz domains in Riemannian manifolds: L^p, Hardy and Hölder type estimates*, to appear in Communications in Analysis and Geometry, (2001).

[37] M. Mitrea and M. Taylor, *Potential theory on Lipschitz domains in Riemannian manifolds: Sobolev-Besov space results and the Poisson problem*, J. Funct. Anal., 176 (2000), 1–79.

[38] M. Mitrea and M. Taylor, *Potential theory on Lipschitz domains in Riemannian manifolds: the case of Hölder continuous metric tensors*, Communications in PDE, Vol. 25 No. 7-8 (2000), 1487–1536.

[39] S. Montgomery-Smith, *private communication.*

[40] J. Peetre, *New Thoughts on Besov Spaces*, Duke Univ. Math. Series, 1976.

[41] J. Pipher and G. Verchota, *Dilation invariant estimates and the boundary Gårding inequality for higher order elliptic operators*, Annals of Math., 142 (1995), 1–38.

[42] L. Rodman, *An introduction to operator polynomials*, Operator theory, advances and applications, 38, Birkhäuser Verlag, 1989.

[43] T. Runst and W. Sickel, *Sobolev Spaces of Fractional Order, Nemytskij Operators, and Nonlinear Partial Differential Operators*, de Gruyter, Berlin, New York, 1996.

[44] Z. Shen, *Boundary value problems for parabolic Lamé systems and a nonstationary linearized system of Navier-Stokes equations in Lipschitz cylinders*, Amer. J. Math., 113 (1991), 293–373.

[45] E. Stein, *Singular Integrals and Differentiability Properties of Functions*, Princeton University Press, 1970.

[46] E. M. Stein, *Topics in harmonic analysis related to Littlewood-Paley theory*, Annals of Math. Studies, Vol. 63, Princeton Univ. Press, Princeton, NJ, 1970.

[47] E. M. Stein and G. Weiss, *Introduction to Fourier Analysis in Euclidean Spaces*, Princeton Univ. Press, Princeton, NJ, 1971.

[48] M. Taylor, *Partial Differential Equations*, Springer-Verlag, 1996.

[49] H. Triebel, *Theory of Function Spaces*, Birkhäuser, Berlin, 1983 (Vol. 2, 1992).

[50] G. Verchota, *Layer potentials and boundary value problems for Laplace's equation in Lipschitz domains*, J. Funct. Anal., Vol. 59 (1984), 572–611.

[51] J. M. Wilson, *A simple proof of atomic decompositions for* $H^p(\mathbb{R}^n)$, $0 < p \leq 1$, Studia Math. 74 (1982), 25–33.

Department of Mathematics, University of Missouri-Columbia, 202 Mathematical Sciences Building, Columbia, MO 65211, USA

E-mail address: `dorina@math.missouri.edu`

Department of Mathematics, University of Missouri-Columbia, 202 Mathematical Sciences Building, Columbia, MO 65211, USA

E-mail address: `marius@math.missouri.edu`

Contemporary Mathematics
Volume **277**, 2001

Growth Theorems and Harnack Inequality for Second Order Parabolic Equations

E. Ferretti and M. V. Safonov

Dedicated to N. V. Krylov, on the occasion of his 60th birthday.

ABSTRACT. A general approach to both divergence (D) and non-divergence (ND) second order parabolic equations is presented, which is based on three growth theorem. These growth theorems look identical in both cases (D) and (ND). They allow to prove the Harnack inequality and other related facts by general arguments, which do not depend on the structure (divergence or non-divergence) of equations. In turn, the growth theorems are established on the grounds of the classical maximum principle and some integral estimates: the energy estimate in the case (D) and the Aleksandrov-Krylov estimate in the case (ND).

Introduction

There are many facts in the theory of second order elliptic and parabolic PDE, which look similar for equations in the divergence and non-divergence forms. These facts include different versions of the Harnack inequality, boundary estimates for solutions, doubling properties for caloric measures, etc. Such kind of results for second order parabolic equations were recently obtained by Fabes, Safonov and Yuan in [**FS**], [**FSY**], [**SY**]. In those papers, different methods were employed for divergence and non-divergence equations. Some of methods, in particular those based on the properties of fundamental solutions, are available only for divergence equations. However, it turns out, the main results there (backward Harnack inequalities, doubling properties for L-caloric measures, and others) can be obtained on the grounds of two principles: the classical maximum principle and the interior Harnack inequality, without using any other specific properties of divergence or non-divergence equations (see [**S98**]).

Unlike the classical maximum principle, the interior Harnack inequality is far from obvious. For elliptic and parabolic equations with measurable coefficients in the *divergence* form, it was proved by Moser in the papers [**M61**],[**M64**]. Many other mathematicians have contributed to this area as well (see monographs

1991 *Mathematics Subject Classification.* Primary 35B05, 35K10; Secondary 35B45, 35J15.
Key words and phrases. Harnack inequality, qualitative properties of solutions, equations with measurable coefficients.
Both authors were supported in part by NSF Grant #9971052.

[F],**[GT]**,**[K]**, **[LE]**, **[L]**,**[LSU]**; papers **[A68]**, **[AS]**, **[DG]**, **[FSt]**, **[I]**, **[LSW]**, **[N]**, **[PE]**, **[SJ]**, **[T]**, and references therein). However, a similar result for *non-divergence* equations was obtained only 15 years later after Moser's papers. This was done by Krylov and Safonov **[KS]**, **[S80]** in 1978-80, who in turn relied on some improved versions of *growth theorems* from the book by Landis **[LE]**. These growth theorems control the behavior of (sub- , super-) solutions of second order elliptic and parabolic equations in terms of the Lebesgue measure of areas in which solutions are positive or negative.

In the present paper, we use growth theorems as a common background for both divergence (D) and non-divergence (ND) equations. We prove three such theorems and derive the interior Harnack inequality as their consequence. For the sake of simplicity, we assume that all functions (coefficients and solutions) are smooth enough. This allows us to treat the cases (D) and (ND) simultaneously. It is easy to get rid of extra smoothness assumptions by means of standard approximation procedures, which are briefly discussed at the end of paper, in Remark 6.1. These procedures, and also minimal smoothness of coefficients and solutions, are different in the cases (D) and (ND). In both cases, it is important to have appropriate estimates for solutions with constants depending only on the prescribed quantities, such as dimension n, parabolicity constant ν, etc., and not depending on "additional" smoothness. We denote such constants $N = N(n, \nu, \cdots)$.

Harnack inequalities have many important applications not only in differential equations, but also in other areas, such as diffusion processes, geometry, etc. For the reader's convenience, we prove most of auxiliary results in their "weak" form, which is sufficient for our purposes. We concentrate on the *parabolic* equations, because the corresponding results for the *elliptic* equations follow automatically.

In Section 1, we introduce our basic assumptions, prove a weak version of the classical *maximum principle* (Theorem 1.4), and formulate the *interior Harnack inequality* (Theorem 1.5). Then we show that even in the one-dimensional case, the Harnack inequality fails for equations of a "joint" structure, which combine both divergence and non-divergence parts. Therefore, it is inevitable that a part of the proof of the Harnack inequality relies on specific properties of equations in the cases (D) and (ND). These are integral estimates in Section 2: the energy estimate in the case (D), and the Aleksandrov-Krylov estimate in the case (ND). In the next three Sections 3,4 and 5, we formulate and prove three growth theorems. Finally, in Section 6 we prove the interior Harnack inequality and describe some possible generalizations.

Basic notations. "$A := B$" or "$B =: A$" is the definition of A by means of the expression B.

$\mathbb{R}^n$ is the n-dimensional Euclidean space , $n \geq 1$, with points $x = (x_1, \cdots, x_n)^t$, where x_i are real numbers. Here the symbol t stands for the transposition of vectors, which indicates that vectors in $\mathbb{R}^n$ are treated as column vectors. For $x = (x_1, \cdots, x_n)^t$ and $y = (y_1, \cdots, y_n)^t$ in $\mathbb{R}^n$, the *scalar product* $(x, y) := \sum x_i y_i$, the *length* of x is $|x| := (x, x)^{1/2}$.

For a Borel set $\Gamma \subset \mathbb{R}^m$, $\partial\Gamma$ is the *boundary* of Γ in $\mathbb{R}^m$, $\overline{\Gamma} := \Gamma \cup \partial\Gamma$ is the *closure* of Γ, $|\Gamma|$ is the m-dimensional Lebesgue measure of Γ. Sometimes we use same notation $|\Gamma|$ for the surface measure of a subset Γ of a smooth surface S.

For real numbers c, we denote $c_+ := \max(c, 0)$, $c_- := \max(-c, 0)$.

1. Maximum principle and Harnack inequality

We discuss basic properties of solutions to second order linear parabolic equations $Lu = 0$ or inequalities, which do not depend on the smoothness of coefficients, and also do not depend on the structure of equations. Namely, we treat simultaneously equations $Lu = 0$ in the *divergence* form, where

$$\text{(D)} \qquad Lu := a_0 u_t - \sum_{i,j=1}^{n} D_i \left(a_{ij} D_j u\right) = a_0 u_t - (D, aDu),$$

and in the *non-divergence* form, where

$$\text{(ND)} \qquad Lu := u_t - \sum_{i,j=1}^{n} a_{ij} D_{ij} u = u_t - (aD, Du).$$

Here all the functions u, a_0 and a_{ij} depend on $X = (x,t) \in \mathbb{R}^{n+1}$, $u_t = \partial u/\partial t$, Du is the vector in $\mathbb{R}^n$ with components $D_i u = \partial u/\partial x_i$, $D_{ij} u = D_i D_j u$, and $a = [a_{ij}]$ is a $n \times n$ matrix with entries $a_{ij} = a_{ij}(X) = a_{ij}(x,t)$, which are defined for all $X = (x,t) \in \mathbb{R}^{n+1}$ and satisfy the *uniform parabolicity* condition

$$\text{(U)} \qquad \min_{\xi \in \mathbb{R}^n, |\xi| = 1} (a\xi, \xi) \geq \nu |\xi|^2, \qquad \|a\|^2 := \sum_{i,j=1}^{n} |a_{ij}|^2 \leq \nu^{-2},$$

with a constant $\nu \in (0,1]$. In addition, in the divergence case we assume that $a_0 = a_0(x)$ does not depend on t and satisfies

$$(\mathrm{U}_0) \qquad \nu \leq a_0(x) \leq \nu^{-1}.$$

In our considerations, functions u, a_0 and a_{ij} are smooth enough, so that all derivatives in Lu are understood in the usual classical sense. More general assumptions require different techniques in the cases (D) and (ND); we discuss these matters in Remark 6.1 at the end of paper. Unless otherwise stated, we always assume that in a given open set Q, functions u and coefficients of L satisfy the following smoothness assumptions.

ASSUMPTIONS 1.1. Function $u \in C^{2,1}(Q)$, i.e. u is continuous together with all the derivatives $D_i u, D_{ij} u, u_t$ in an open set $Q \subset \mathbb{R}^{n+1}$. Functions $a_0 = a_0(x)$ and a_{ij} are continuous and satisfy (U), (U_0) with a constant $\nu \in (0,1]$. In addition, in the case (D), a_{ij} have continuous derivatives $D_k a_{ij}$ for $i, j, k = 1, \cdots, n$.

It is easy to see these assumptions guarantee that Lu in continuous in Q.

REMARK 1.2. In the non-divergence case (ND), the additional coefficient a_0 in front of u_t is not needed, because this case is easily reduced to $a_0 = 1$: one can simply divide by a_0 and replace the matrix a by $\tilde{a} = a_0^{-1} a$, which satisfy (U) with $\tilde{\nu} = \nu^2$ instead of ν. Moreover, in the case (ND) we can always replace a_{ij} by $\frac{1}{2}(a_{ij} + a_{ji})$, so that the matrix a becomes symmetric. Furthermore, under parallel translations in $\mathbb{R}^{n+1}$ and rotations with respect to the space variable $x \in \mathbb{R}^n$, the equations $Lu = 0$ are transformed to similar equations with new coefficients, which satisfy the conditions (U) and (U_0) with the same constant $\nu \in (0,1]$.

For formulation of our results, we need some standard definitions and notations.

DEFINITION 1.3. Let Q be an open connected set in $\mathbb{R}^{n+1}$, $n \geq 1$. The *parabolic boundary* $\partial_p Q$ of Q is the set of all points $X_0 = (x_0, t_0) \in \partial Q$, such that there exists a continuous function $x = x(t)$ on an interval $(0, \delta)$ with values in $\mathbb{R}^n$, such that

$x(t_0) = x_0$ and $(x(t), t) \in Q$ for all $t \in (t_0, t_0 + \delta)$. Here $x = x(t)$ and $\delta > 0$ depend on X_0.

In particular, for cylinders $Q = \Omega \times (0, T)$, the parabolic boundary $\partial_p Q := (\partial_x Q) \cup (\partial_t Q) \cup (\partial_{xt} Q)$, where

$$\partial_x Q := (\partial\Omega) \times (0, T), \ \partial_t Q := \Omega \times \{0\}, \ \partial_{xt} Q := (\partial\Omega) \times \{0\}.$$

We will also use a notation for "standard" parabolic cylinders: for $Y = (y, s)$ and $r > 0$,

$$C_r(Y) := B_r(y) \times \left(s - r^2, s\right), \text{ where } B_r(y) := \{x \in \mathbb{R}^n : |x - y| < r\}.$$

The following *maximum principle* is well know. We give its short proof, which uses our simplifying assumptions.

THEOREM 1.4 (Maximum Principle). *Let Q be a bounded open set in $\mathbb{R}^{n+1}$, and let a function $u \in C^{2,1}\left(\overline{Q} \setminus \partial_p Q\right) \cap C\left(\overline{Q}\right)$ satisfy the inequality $Lu \leq 0$ in Q. Then*

$$\sup_Q u = \sup_{\partial_p Q} u. \tag{1.1}$$

PROOF. Since a_{ij} are smooth, we can write

$$Lu = a_0 u_t - (aD, Du) + (b, Du) \qquad \text{in} \quad Q,$$

where the vector function b has components $b_j = \sum_i D_i a_{ij}$ in the case (D), and $a_0 = 1, b = 0$ in the case (ND). Suppose (1.1) is not true. Replacing u by $u - const$, we may assume

$$\sup_{\partial_p Q} u < 0 < \sup_Q u = u(Y) \qquad \text{for some} \quad Y = (y, s) \in \overline{Q} \setminus \overline{\partial_p Q}.$$

In addition, replacing u by $u - \varepsilon t$ with a small constant $\varepsilon > 0$, we have $Lu(Y) < 0$. Moreover, replacing a_{ij} by $\frac{1}{2}(a_{ij} + a_{ji})$, we can assume that the matrix $a = [a_{ij}]$ is symmetric. Then by rotating the coordinate axes, we can reduce the matrix of coefficients to the diagonal matrix $diag[\lambda_1, \ldots, \lambda_n]$, where $\lambda_i \in \left[\nu, \nu^{-1}\right]$ are eigenvalues of a. Further, since $Y = (y, s) \in \overline{Q} \setminus \overline{\partial_p Q}$, we also have $\overline{C_r(Y)} \subset \overline{Q} \setminus \overline{\partial_p Q}$ for some small $r > 0$, so that

$$u \in C^{2,1}\left(\overline{C_r(Y)}\right) \qquad \text{and} \qquad \sup_{C_r(Y)} u = u(Y).$$

Since $u(\cdot, s)$ attains its maximum at y, we have $D_i(Y) = 0, \ D_{ii}(Y) \leq 0$ for all i, and similarly, $u_t(Y) \geq 0$. Combining these relations, we get

$$Lu(Y) = a_0 u_t(Y) - \sum_i \lambda_i D_{ii}(Y) + \sum_i b_i D_i(Y) \geq 0.$$

On the other hand, by our construction $Lu(Y) < 0$. This contradictions proves the desired equality (1.1). □

In the next theorem, we formulate the *interior Harnack inequality*. The proof of this theorem will be given in Section 6.

THEOREM 1.5 (Interior Harnack Inequality). *Let u and the coefficients of L satisfy Assumptions 1.1 in a cylinder $C_{2r} = C_{2r}(Y)$, where $Y = (y, s) \in \mathbb{R}^{n+1}, r > 0$, and let u satisfy $u \geq 0$, $Lu = 0$ in C_{2r}. Then*

$$\sup_{C^0} u \leq N \inf_{C_r} u, \tag{1.2}$$

where $C^0 := B_r(y) \times (s-3r^2, s-2r^2)$ *and the constant* $N = N(n,\nu)$.

For second order elliptic equations, this theorem is reduced to the following

COROLLARY 1.6. *Let* $u = u(x)$ *and* $a_{ij} = a_{ij}(x)$ *satisfy Assumptions 1.1 in a ball* $B_{2r} = B_{2r}(y)$, *where* $y \in \mathbb{R}^n, r > 0$, *and let* u *satisfy*

$$u \geq 0 \quad \text{and} \quad (D) \quad (D, aDu) = 0 \quad \text{or} \quad (ND) \quad (aD, Du) = 0 \quad \text{in} \quad B_{2r}.$$

Then

$$\sup_{B_r} u \leq N \inf_{B_r} u, \tag{1.3}$$

where $N = N(n,\nu)$.

The proof of the maximum principle, Theorem 1.4, with minor changes remains valid for more general "united" equations $u_t - (D, a_1 Du) - (a_2 D, Du) = 0$, which include both divergence and non-divergence equations. However, for such equations the Harnack inequality (1.3) fails even in the one-dimensional elliptic case, as the following example shows.

EXAMPLE 1.7. For arbitrary small $\varepsilon \in \left(0, \frac{1}{2}\right)$, we will construct smooth functions a_1, a_2 and u on $[-1,1]$, such that $|a_{1,2} - 1| \leq \varepsilon$, $0 \leq u \leq 1$,

$$u' > 0, \quad (a_1 u')' + a_2 u'' = 0 \quad \text{in} \quad (-1,1), \tag{1.4}$$

and

$$0 \leq u \leq \varepsilon \quad \text{on} \quad [-1,-\varepsilon], \qquad 1-\varepsilon \leq u \leq 1 \quad \text{on} \quad [\varepsilon, 1]. \tag{1.5}$$

Obviously, from these properties it follows that the estimate (1.3) fails for $y = 0$ and $r = \frac{1}{2}$.

For fixed $\varepsilon \in \left(0, \frac{1}{2}\right)$, take an arbitrary smooth function $\eta = \eta(x)$ on $[-1,0]$, satisfying the properties $|\eta| \leq \frac{\varepsilon}{2}$, $|\eta'| \leq 1$ on $[-1,0]$;

$$\eta > 0 \quad \text{in} \quad \left(-\varepsilon, -\frac{\varepsilon}{2}\right), \quad \text{and} \quad \eta(x) \equiv 0 \quad \text{on} \quad [-1,-\varepsilon] \cup \left[-\frac{\varepsilon}{2}, 0\right].$$

For a constant $\omega > 0$ to be chosen later, we first define the functions

$$a_1(x) = 1 - \eta \sin \omega x, \quad a_2(x) = 1 + \eta \cdot (\sin \omega x - \cos \omega x) \qquad \text{on} \quad [-1,0],$$

and then extend them as even functions to the interval $[-1,1]$. We can rewrite the equality (1.4) in the form

$$(a_1 + a_2) u'' + a_1' u' = 0, \qquad \text{or} \quad f := (\ln u')' = \frac{u''}{u'} = -\frac{a_1'}{a_1 + a_2}.$$

Therefore, the function

$$u(x) = \frac{1}{A} \int_{-1}^{x} e^{F(z)} dz, \qquad \text{where} \quad F(x) = \int_{-1}^{x} f(z)\,dz, \quad A = \int_{-1}^{1} e^{F(z)} dz > 0,$$

satisfies (1.4) and $u(-1) = 0, u(1) = 1$. Now it remains to show that for large $\omega > 0$ (depending on ε), this function satisfies (1.5) as well.

Since $a_{1,2}$ are even functions, $a_{1,2}(-x) \equiv a_{1,2}(x)$, we also have

$$f(-x) \equiv -f(x), \quad F(-x) \equiv F(x), \quad u(-x) + u(x) \equiv 1 \qquad \text{on} \quad [-1,1].$$

By the last equality here, $0 \le u \le \varepsilon$ on $[-1,-\varepsilon]$ if and only if $1-\varepsilon \le u \le 1$ on $[\varepsilon, 1]$, so that it suffices to prove the estimate on the interval $[-1,-\varepsilon]$. We know that

$$f(x) = \frac{(\eta \sin \omega x)'}{2-\eta \cos \omega x} \qquad \text{on} \quad [-1,0].$$

Integrating by parts and using the properties $|\eta| \le \frac{\varepsilon}{2} < \frac{1}{4}$, $|\eta'| \le 1$, we estimate

$$\begin{aligned}
F(x) &= \int_{-1}^{x} f(z)\,dz = \int_{-1}^{x} \frac{(\eta \sin \omega z)'}{2-\eta \cos \omega z} dz \\
&= -\int_{-1}^{x} \eta \sin \omega z \left(\frac{1}{2-\eta \cos \omega z}\right)' dz + O(\varepsilon) \\
&= \omega \int_{-1}^{x} \left(\frac{\eta \sin \omega z}{2-\eta \cos \omega z}\right)^2 dz + O(\varepsilon) \\
&\ge \frac{\omega}{12} \int_{-1}^{x} \eta^2 \sin^2 \omega z\, dz + O(\varepsilon) \\
&= \frac{\omega}{12} \int_{-1}^{x} \eta^2 dz - \frac{1}{12} \int_{-1}^{x} \eta^2 (\sin 2\omega z)'\, dz + O(\varepsilon) \\
&= \frac{\omega}{12} \int_{-1}^{x} \eta^2 dz + O(\varepsilon) \qquad \text{on} \quad [-1,0].
\end{aligned}$$

Since $\eta > 0$ in $\left(-\varepsilon, -\frac{\varepsilon}{2}\right)$, the values of $F(x)$ in this interval, and hence the constant $A > 0$, become arbitrary large for large $\omega > 0$. Choose ω such that $A \ge \varepsilon^{-1}$. Then the desired estimate $0 \le u \le \varepsilon$ on $[-1,-\varepsilon]$ follows immediately, because

$$\eta \equiv 0, \quad f \equiv 0, \quad F \equiv 0, \quad u(x) = \frac{1}{A}(x+1) \qquad \text{on} \quad [-1,-\varepsilon].$$

2. Integral estimates

In this sections, we prove basic integral estimates for solutions of second order parabolic equations: the energy estimate in the divergence case (D) and the Aleksandrov-Krylov estimate in the non-divergence case (ND). For formulation of these results, we need further notations.

Let Ω be an open set in $\mathbb{R}^n$. For $0 < p < \infty$, $L^p(\Omega)$ denotes the linear space of measurable functions with the finite

$$(2.1) \qquad \|u\|_p = \|u\|_{p,\Omega} := \left(\int_\Omega |u|^p\, dx\right)^{1/p}.$$

For $1 \le p < \infty$, $L^p(\Omega)$ is a Banach space with the norm $\|\cdot\|_p$. For $p = \infty$, $L^\infty(\Omega)$ is the Banach space of bounded functions on Ω with the norm

$$(2.2) \qquad \|u\|_\infty = \|u\|_{\infty,\Omega} := \sup_\Omega |u|.$$

We will also use a similar notation $L^p(Q)$ for functions $u(X) = u(x,t)$ on an open set $Q \subset \mathbb{R}^{n+1}$. These notations make sense for vector functions u as well, with the understanding that $|u|$ in (2.1) and (2.2) stands for the length of vector $u = u(x)$.

Let Q be an open set in $\mathbb{R}^{n+1}$. For functions $u = u(x,t)$ on Q, which have derivatives $D_i u$, we denote

$$(2.3) \qquad \|u\|^2_{V(Q)} := \int_Q |Du|^2 \, dX + \sup_t \int_{Q(t)} |u(x,t)|^2 \, dx,$$

where $Q(t) := \{x \in \mathbb{R}^n : (x,t) \in Q\}$.

We start with a simple lemma, which holds in both cases (D) and (ND). Its proof is straightforward.

LEMMA 2.1. *Let Q be an open set in $\mathbb{R}^{n+1}$. Then for arbitrary functions $u \in C^{2,1}(Q)$ and $G \in C^2(\mathbb{R}^1)$, the function $v := G(u)$ belongs to $C^{2,1}(Q)$ and satisfies*

$$Dv = G'(u)\, Du, \quad Lv = G'(u)\, Lu - G''(u)\,(aDu, Du) \quad \text{in} \quad Q.$$

In particular, if $u \geq 0, Lu \leq 0$ in Q, and $G, G', G'' \geq 0$, then $v \geq 0, Lv \leq 0$ in Q.

The following theorem is a modification of well known *energy estimates* (see [**LSU**], Sec. 3.2; [**L**], Sec. 6.1). It holds for equations in the divergence form (D).

THEOREM 2.2 (Energy Estimate). *Let u be a function in $C^{2,1}(C_2)$, where $C_2 = C_2(Y) = B_2(y) \times (s-4, s)$, $Y = (y,s) \in \mathbb{R}^{n+1}$, and let*

$$(2.4) \qquad Lu := a_0 u_t - (D, aDu) = f_0 + (D, f) = f_0 + \sum_{i=1}^n D_i f_i$$

in C_2, with functions $f_0 \in C(C_2)$, $f = (f_1, \cdots, f_n)^t$, $Df_i \in C(C_2)$ for $i = 1, \cdots, n$. Take a standard [1] smooth cut-off function η, such that

$$(2.5) \qquad 0 \leq \eta \leq 1 \quad \text{on } \overline{C_2}, \quad \eta \equiv 1 \quad \text{on} \quad \overline{C_1(Y)}, \quad \text{and} \quad \eta = 0 \quad \text{near} \quad \partial_p C_2.$$

Then

$$(2.6) \qquad \|\eta u\|^2_{V(C_2)} \leq NF^2, \quad \text{where} \quad F^2 := \int_{C_2} \left(u^2 + f_0^2 + |f|^2\right) dX,$$

and the constant $N = N(n, \nu)$. This estimate remains true if instead of the equality (2.4) in C_2, we impose two inequalities

$$(2.7) \qquad u \geq 0, \quad Lu \leq f_0 + (D, f).$$

PROOF. Consider the function

$$I(t) := \int a_0 \eta^2 u^2(x,t)\, dx, \quad s - 4 \leq t \leq s,$$

where the integral is taken over $B_2(y)$. From (2.4) or (2.7) it follows

$$\begin{aligned} I'(t) &= \int \left(a_0\eta^2\right)_t \cdot u^2 dx + \int 2\eta^2 u \cdot a_0 u_t \, dx \\ &\leq N \int u^2 dx + 2 \int \eta^2 u \left[(D, aDu) + f_0 + (D, f)\right] dx. \end{aligned}$$

[1]By standard functions we mean smooth functions with derivatives bounded by constants depending only on the order of derivative and the dimension n.

Then we integrate by parts and use the estimates $2uf_0 \le u^2 + f_0^2$, and for $0 < \varepsilon < 1$,

$$\begin{aligned} -2\left(D\left(\eta^2 u\right), f\right) &= -2\eta\left(D\left(\eta u\right), f\right) - 2\eta u\left(D\eta, f\right) \\ &\le \varepsilon\left|D\left(\eta u\right)\right|^2 + \varepsilon^{-1}\left|f\right|^2 + N\left(u^2 + \left|f\right|^2\right), \\ -2\left(D\left(\eta^2 u\right), aDu\right) &= -2\left(D\left(\eta u\right) + uD\eta, a\eta Du\right) \\ &= -2\left(D\left(\eta u\right) + uD\eta, aD\left(\eta u\right) - uaD\eta\right) \\ &\le -2\nu\left|D\left(\eta u\right)\right|^2 + N\left|u\right| \cdot \left|D\left(\eta u\right)\right| + Nu^2 \\ &\le \left(\varepsilon - 2\nu\right)\left|D\left(\eta u\right)\right|^2 + N\varepsilon^{-1}u^2. \end{aligned}$$

This gives us

$$I'(t) \le 2\left(\varepsilon - \nu\right)\int \left|D\left(\eta u\right)\right|^2 dx + N\varepsilon^{-1}\int \left(u^2 + f_0^2 + \left|f\right|^2\right) dx, \quad 0 < \varepsilon < 1.$$

For $\varepsilon = \frac{\nu}{2}$, this implies

$$\nu\int \left|D\left(\eta u\right)\right|^2 dx + I'(t) \le N\int \left(u^2 + f_0^2 + \left|f\right|^2\right) dx, \qquad s - 4 \le t \le s.$$

Integrating with respect to t and using Fubini's theorem, we conclude

$$\nu\int_{C_2} \left|D\left(\eta u\right)\right|^2 dX \le NF^2, \qquad \sup_{s-4<t<s} I(t) \le NF^2.$$

Since $a_0 \ge \nu > 0$, we also have

$$\nu\int_{B_2(y)} \left(\eta u\right)^2(x,t)\, dx \le I(t).$$

From these estimates it follows (2.6) with a constant $N = N(n,\nu)$. □

COROLLARY 2.3. *Let a function $u \in C^{2,1}\left(\overline{Q}\right)$, where Q is an open set in $\mathbb{R}^{n+1}$, and let*

$$\begin{aligned} u &\ge 0, \quad Lu := a_0 u_t - \left(D, aDu\right) \le 0 \quad \text{in} \quad Q; \\ u &= 0 \quad \text{on} \quad \left(\partial Q\right) \cap C_2, \end{aligned}$$

where $C_2 = C_2(Y) = B_2(y) \times (s-4, s)$, $Y = (y,s) \in \mathbb{R}^{n+1}$. Let η be a standard smooth cut-off function η satisfying the conditions (2.4) in the cylinder $C_2 = C_2(Y)$. Then

$$\left\|\eta u\right\|^2_{V(Q\cap C_2)} \le NF^2, \quad \text{where} \quad F^2 := \int_{Q\cap C_2} u^2 dX, \tag{2.8}$$

and the constant $N = N(n,\nu)$.

PROOF. We approximate u by smooth functions $u_\varepsilon = G_\varepsilon(u)$ for small $\varepsilon > 0$, where G_ε are smooth functions such that $G_\varepsilon, G'_\varepsilon, G''_\varepsilon \ge 0$ on $\mathbb{R}^1$, $G_\varepsilon(u) \equiv 0$ for $u \le \varepsilon$, and $G'_\varepsilon(u) \equiv 1$ for $u \ge 2\varepsilon$. Then

$$u - 2\varepsilon \le u_\varepsilon \le \max\left(u - \varepsilon, 0\right) \quad \text{on} \quad \overline{Q}, \quad Du_\varepsilon = Du \quad \text{on} \quad \left\{u \ge 2\varepsilon\right\}. \tag{2.9}$$

Moreover, since $u_\varepsilon \equiv 0$ near $(\partial Q) \cap C_2(Y)$, we can extend $u_\varepsilon \equiv 0$ on $C_2 \setminus Q$, the extended functions $u_\varepsilon \in C^\infty(C_2)$, and by Lemma 2.1, $u_\varepsilon \geq 0, Lu_\varepsilon \leq 0$ in C_2. From Theorem 2.2 it follows that the functions u_ε satisfy the estimate

$$||u_\varepsilon||_{V(C_2)} \leq N(n,\nu) \int_{C_2} u_\varepsilon^2 dX, \quad \varepsilon > 0.$$

Taking the limit as $\varepsilon \to 0^+$ and using (2.9), we get the estimate (2.8). □

For solutions of second order elliptic and parabolic equations $Lu = f$ with measurable coefficients in the non-divergence form (ND), there are pointwise estimates through the norm $||f||_p$, where $p = n$ in the elliptic case, and $p = n+1$ in the parabolic case. In the elliptic case, such estimates were established by Aleksandrov [**A**] (see also [**GT**], Cf. 9, for further discussion and references to related works by Bakel'man, Pucci and other mathematicians). A similar estimate for solutions of parabolic equations was obtained by Krylov [**K76**], who used a completely different method in his proof. Later Tso [**TK**] found a simpler proof of Krylov's result by further adjustment of the original Aleksandrov's method. In the next theorem, we present the Aleksandrov-Krylov estimate, in its simplified form. In the proof, we basically follow Tso [**TK**].

THEOREM 2.4. *Let a function $u \in C^{2,1}(\overline{Q})$, where Q is an open subset of $C_1 = C_1(Y_0) \subset \mathbb{R}^{n+1}$, and let*

$$Lu = u_t - (aD, Du) \leq f \quad \text{in} \quad Q, \qquad u \leq 0 \quad \text{on} \quad \partial_p Q, \tag{2.10}$$

where $f \in L^{n+1}(Q)$. Then

$$M := \sup_Q u \leq NF, \quad \text{where} \quad F := ||f||_{n+1,Q}, \tag{2.11}$$

and the constant $N = N(n,\nu)$.

PROOF. Obviously, it suffices to consider the case $M > 0$. Then we have $M = u(X_0)$ for some $X_0 = (x_0, t_0) \in \overline{Q} \setminus \partial_p Q$. We define the *upper contact set* $\Gamma = \Gamma_u$ of u as the set of all $X = (x,t) \in Q$ such that

$$u(X) + (y - x, Du(X)) \geq u(Y)$$

for all $Y = (y,s) \in Q$ with $s \leq t$. Introduce the mapping $\Phi : \Gamma \to \mathbb{R}^{n+1}$ by formulas

$$\Phi = (p,h), \qquad p = Du, \qquad h = u - (x - x_0, Du).$$

Since $D_j h = -(x - x_0, DD_j u)$ and $h_t = u_t - (x - x_0, Du_t)$, the Jacobian matrix [2]

$$J_\Phi = \frac{\partial \Phi}{\partial X} = \frac{\partial(p,h)}{\partial(x,t)} = \begin{bmatrix} D_{ij}u & D_i u_t \\ (x_0 - x, DD_j u) & u_t + (x_0 - x, Du_t) \end{bmatrix}$$

has determinant $\det J_\Phi = u_t \det[D_{ij}u] = u_t \det D^2 u$.

We claim that $\Phi(\Gamma)$ contains the cylinder

$$C = \left\{(p,h) \in \mathbb{R}^n : |p| < \frac{1}{4}M, \ \frac{1}{2}M < h < M\right\} \subset \Phi(\Gamma).$$

[2] Actually the calculation of Du_t in this matrix requires higher smoothness of u. However, by a simple approximation argument (see Remark 6.1) the estimate (2.11) is extended from functions $u \in C^\infty(\overline{Q})$ to more general class $W^{2,1}_{n+1,loc}(Q) \cap C(\overline{Q})$.

Indeed, fix $(p,h) \in C$ and consider the function $l(x) = (x - x_0, p) + h$. Since $Q \subset C_1$, we have

$$|x - x_0| < 2,\ l(x) > 0 \geq u(x,t) \quad \text{for} \quad (x,t) \in \partial_p Q, \quad l(x_0) = h < M = u(x_0, t_0).$$

Therefore, the set $Q \cap \{u = v\}$ is not empty. Choose a point $X = (x,t)$ in this set, which has the minimal possible coordinate t. Then

$$X = (x,t) \in \Gamma,\ u(X) = l(x),\ Du(X) = Dl(x) = p,$$

and

$$\Phi(X) = (Du(X), u(X) - (x - x_0, Du(X))) = (p,h).$$

Since (p,h) is an arbitrary point in C, we have proved $C \subset \Phi(\Gamma)$. This implies the inequality for the Lebesgue measures:

$$\frac{1}{N} M^{n+1} = |C| \leq |\Phi(\Gamma)| = \int_\Gamma |\det J_\Phi|\, dX = \int_\Gamma \left|u_t \det D^2 u\right| dX. \tag{2.12}$$

From geometrical properties of Γ, we have $u_t \geq 0$ and $D_{ii}u \leq 0$ for all i on Γ. Further, for any fixed point $X \in \Gamma$, we can rotate the coordinate axes in such a way that the matrix $D^2u = [D_{ij}u]$ becomes diagonal at X. In the new coordinates,

$$\det D^2 u = \prod_{i=1}^n D_{ii}u, \quad Lu = u_t - \sum_{i=1}^n a_{ii} D_{ii} u \geq \nu (u_t - \Delta u) \geq 0 \quad \text{at } X.$$

Using the inequality between the geometrical mean and the arithmetical mean of $n+1$ nonnegative numbers $u_t, -D_{11}u, \ldots, -D_{nn}u$, we get

$$(n+1)\left|u_t \det D^2 u\right|^{\frac{1}{n+1}} \leq u_t - \Delta u \leq \nu^{-1} Lu \leq \nu^{-1} f \quad \text{on} \quad \Gamma.$$

Now the desired estimate (2.11) follows from (2.12). □

3. First growth theorem

In this section, we formulate and prove the first of three *growth theorems*, Theorem 3.3. This theorem, as well as the second and third growth theorems, Theorems 4.2 and 5.3 have exactly same formulation in both divergence (D) and non-divergence (ND) cases. Roughly speaking, they claim that from the inequality $Lu \leq 0$ in a cylinder $C_r(Y)$, and some additional information on the set $\{u \leq 0\}$, it follows

$$M_{r/2}(Y) \leq \beta \cdot M_r(Y), \quad \text{where} \quad M_r(Y) := \sup_{C_r(Y)} u_+, \tag{3.1}$$

with a constant $\beta \in (0,1)$. It is convenient to have a special notation for a class of functions satisfying this estimate.

DEFINITION 3.1. Let $Y = (y,s) \in \mathbb{R}^{n+1}$, $r > 0$, and $\beta = const \in (0,1)$ be fixed. Denote $\mathcal{M}(\beta, Y, r)$ the class of all functions u which are defined on $C_r(Y)$ and satisfy the estimate (3.1).

Obviously, this estimate imposes some restriction only on the positive part $u_+ = \max(u,0)$ of u; the class $\mathcal{M}(\beta, Y, r)$ automatically contains all functions u satisfying $u \leq 0$ on $C_r(Y)$.

In the divergence case (D), in addition to the energy estimate in Theorem 2.2, we need the following technical lemma.

LEMMA 3.2. *Let Ω be a bounded open set in $\mathbb{R}^n$, and let u be a continuously differentiable function on $\overline{\Omega}$ which vanishes on the boundary $\partial\Omega$ of Ω. Then*

$$\|u\|_1 := \int_\Omega |u|\, dx \le N\, |\Omega|^{1/n} \int_\Omega |Du|\, dx, \tag{3.2}$$

where the constant $N = N(n) > 0$.

PROOF. This inequality with $N = 1$ follows immediately from Hölder's inequality and the Gagliardo-Nirenberg-Sobolev imbedding theorem (see [**GT**], Sec. 7.7, or [**LSU**], Sec. 2.2). Indeed, since (3.2) is invariant under rescaling $x \to const \cdot x$, we may assume $|\Omega| = 1$, and then

$$\|u\|_1 \le \|u\|_{\frac{n}{n-1}} \le \int_\Omega |Du|\, dx.$$

For completeness, we give another elementary proof of (3.2). We assume

$$u \equiv 0, \quad Du \equiv 0 \quad \text{on} \quad \mathbb{R}^n \setminus \Omega.$$

Then for arbitrary $y \in \mathbb{R}^n$ and θ in the unit sphere $S := \{|\theta| = 1\} \subset \mathbb{R}^n$,

$$\begin{aligned} u(y) &= -\int_0^\infty \frac{d}{dr} u(y - r\theta)\, dr = \int_0^\infty (\theta, Du(y - r\theta))\, dr, \\ |u(y)| &\le \int_0^\infty |Du(y - r\theta)|\, dr. \end{aligned}$$

Integrating over S with respect to θ, considering r and θ as radial and angular coordinates of $x = r\theta \in \mathbb{R}^n$, and applying Fubini's theorem, we obtain

$$\begin{aligned} |u(y)| &\le \frac{1}{|S|} \int_S \int_0^\infty |Du(y - r\theta)|\, dr\, d\theta = \frac{1}{|S|} \int_{\mathbb{R}^n} |Du(y - x)| \cdot |x|^{1-n}\, dx \\ &= \frac{1}{|S|} \int_\Omega |Du(x)| \cdot |y - x|^{1-n}\, dx, \end{aligned}$$

$$\int_\Omega |u|\, dy = \frac{1}{|S|} \int_\Omega \int_\Omega |Du(x)| \cdot |y - x|^{1-n}\, dx\, dy = \int_\Omega |Du(x)| \cdot I(x)\, dx, \tag{3.3}$$

where

$$I(x) := \frac{1}{|S|} \int_\Omega |y - x|^{1-n}\, dy.$$

Now it remains to show that $I(x) \le N\, |\Omega|^{1/n}$ for all x. For the proof of this estimate, choose $\rho > 0$ such that $|B_\rho| = |\Omega|$. Then obviously $|\Omega \setminus B_\rho(x)| = |B_\rho(x) \setminus \Omega|$. Moreover,

$$\sup_{\Omega \setminus B_\rho(x)} |y - x|^{1-n} \le \rho^{1-n} \le \inf_{B_\rho(x) \setminus \Omega} |y - x|^{1-n}.$$

Comparing the integrals, we get

$$\int_{\Omega} |y-x|^{1-n}\,dy \le \int_{B_\rho(x)} |y-x|^{1-n}\,dy = \int_{B_\rho(0)} |y|^{1-n}\,dy$$
$$= \int_S \int_0^\rho r^{1-n}\cdot r^{n-1} dr\,d\theta = |S|\cdot\rho,$$

hence

$$I(x) \le \rho = |B_1|^{-1/n}\cdot|B_\rho|^{1/n} = N\,|\Omega|^{1/n},$$

where $N = N(n) = |B_1|^{-1/n}$. From the last estimate and (3.3), the estimate (3.2) follows. □

THEOREM 3.3 (First Growth Theorem). *Let a function* $u \in C^{2,1}\left(\overline{C_r}\right)$, *where* $C_r := C_r(Y)$, $Y = (y,s) \in \mathbb{R}^{n+1}$, $r > 0$, *and let* $Lu \le 0$ *in* C_r. *In addition, suppose*

$$|\{u>0\}\cap C_r| \le \mu_1\cdot|C_r|, \quad \textit{where} \quad \mu_1 = const \in (0,1]. \tag{3.4}$$

Then $u \in \mathcal{M}(\beta_1, Y, r)$ *with a constant* $\beta_1 = \beta_1(n,\nu,\mu_1) \in (0,1]$, *such that* $\beta_1 \to 0^+$ *as* $\mu_1 \to 0^+$.

PROOF. By Definition 3.1 of classes $\mathcal{M}$, we have to prove the estimate (3.1) with $\beta = \beta_1 \to 0^+$ as $\mu_1 \to 0^+$. Note that instead of (3.1), it suffices to prove a weaker estimate

$$u(Y) \le \beta_1\cdot M_r(Y), \quad \text{where} \quad M_r(Y) := \sup_{C_r(Y)} u_+. \tag{3.5}$$

Indeed, for arbitrary $Z \in C_{r/2}(Y)$ we have $C_{r/2}(Z) \subset C_r(Y) =: C_r$, and (3.4) implies

$$\left|\{u>0\}\cap C_{r/2}(Z)\right| \le |\{u>0\}\cap C_r| \le \mu_1\cdot|C_r| = 2^{n+2}\mu_1\cdot\left|C_{r/2}\right|.$$

Hence we can use (3.5) with Z and $2^{n+2}\mu_1$ instead of Y and μ_1, correspondingly. Then $u(Z) \le \beta_1\cdot M_r(Z) \le \beta_1\cdot M_r(Y)$, and since Z is an arbitrary point in $C_{r/2}(Y)$, the estimate (3.1) follows. For the proof of (3.5), we consider separately cases (D) and (ND).

Case (D). Using rescaling $x \to cx, t \to c^2 t$ with an appropriate constant $c > 0$, and replacing u by $const\cdot u$, we reduce the proof to the case $r = 2$ and $M_2(Y) = 1$. We set $Q := \{u>0\}\cap C_2$, and for $t \in I := (s-4, s)$,

$$f(t) := |Q(t)|, \quad g(t) := \int_{Q(t)} |Dw(x,t)|^2\,dx, \quad W(t) := \int_{Q(t)} |w(x,t)|\,dx,$$

where $Q(t) := \{x \in \mathbb{R}^n : (x,t) \in Q\}$. By Fubini's theorem and (3.4),

$$\|f\|_{1,I} = \int_I |Q(t)|\,dt = |Q| \le N\mu_1, \quad \text{where} \quad N = N(n) = |C_2|. \tag{3.6}$$

Moreover, since $u \le 1$ in Q, by Corollary 2.3 the function $w := \eta u$ satisfies

$$\|w\|^2_{V(Q)} \le N\int_Q u^2 dX \le N\cdot|Q| \le N\mu_1,$$

with different constants $N = N(n, \nu)$. This estimate includes

$$(3.7) \qquad ||g||_{1,I} = \int\limits_I \int\limits_{Q(t)} |Dw(x,t)|^2 \, dx \, dt = \int\limits_Q |Dw|^2 \, dX \le N\mu_1$$

and

$$\int\limits_{Q(t)} w^2(x,t) \, dx \le N\mu_1 \quad \text{for} \quad t \in I.$$

Using Hölder's inequality, we get

$$W^2(t) = ||w(\cdot,t)||^2_{1,Q(t)} \le |Q(t)| \cdot \int\limits_{Q(t)} w^2(x,t) \, dx \le N\mu_1 f(t).$$

On the other hand, by Lemma 3.2 and Hölder's inequality,

$$W^2(t) \le N |Q(t)|^{\frac{2}{n}} ||Dw(\cdot,t)||^2_{1,Q(t)} \le N f^{\frac{2}{n}+1}(t) \cdot g(t).$$

These two estimates allow us to write

$$W(t) = W^{2\alpha}(t) \cdot W^{1-2\alpha}(t) \le N\mu_1^\alpha f^{\alpha_1}(t) \cdot g^{\alpha_2}(t), \quad \alpha_1 := \frac{1}{2} + \frac{1-2\alpha}{n}, \ \alpha_2 := \frac{1}{2} - \alpha,$$

with an arbitrary $\alpha \in \left(0, \frac{1}{2}\right)$. We take $\alpha := \frac{1}{n+2}$ in order to guarantee the equality $\alpha_1 + \alpha_2 = 1$. Using Fubini's theorem, Hölder's inequality, and estimates (3.6), (3.7), we obtain

$$\int\limits_Q w(X) \, dX = \int\limits_I W(t) \, dt \le N\mu_1^\alpha \int\limits_I f^{\alpha_1} g^{\alpha_2} \, dt \le N\mu_1^\alpha \cdot ||f||^{\alpha_1}_{1,I} \cdot ||g||^{\alpha_2}_{1,I} \le N\mu_1^{1+\alpha}.$$

Replacing μ_1 in s (3.4) by a smaller constant if necessary, we may assume $|Q| = |\{u > 0\} \cap C_2| = \mu_1 \cdot |C_2|$. Since $w = \eta u \equiv u$ on $Q \cap C_1$, the previous estimate implies

$$\int\limits_{C_1} u_+ \, dX \le \int\limits_Q w \, dX \le N \cdot |\{u > 0\} \cap C_2|^{1+\alpha}, \quad \alpha = \frac{1}{n+2} > 0.$$

The previous arguments remain valid for the functions $u - h$, $h = const \ge 0$, with Q being replaced by $\{u > h\} \cap C_2$. Then the last estimate has the form

$$\int_{C_1} (u-h)_+ \, dX \le N \cdot |\{u > h\} \cap C_2|^{1+\alpha}, \qquad h \ge 0.$$

Using rescaling $x \to \rho x$, $t \to \rho^2 t$, we can rewrite this estimate for arbitrary cylinder $C_{2\rho}$ instead of C_2. Since $|C_\rho| = |B_1| \cdot \rho^{n+2}$ and $\alpha = \frac{1}{n+2}$, after cancellations we arrive at

$$(3.8) \qquad \int_{C_\rho} (u-h)_+ \, dX \le \frac{N_0}{\rho} \cdot |\{u > h\} \cap C_{2\rho}|^{1+\alpha}, \quad 0 < \rho \le 1, \ h \ge 0,$$

where $N_0 = N_0(n, \nu) > 0$.

By our assumption $M_2(Y) = 1$, we can rewrite (3.5) as $u(Y) \le \beta_1$. In order to prove this estimate (with a small $\beta_1 > 0$ for small $\mu_1 > 0$) it suffices to show that the estimate

$$(3.9) \qquad m(\rho) := |\{u > (1-\rho)\beta_1\} \cap C_{2\rho}(Y)| \le \mu_1 \rho^\gamma |C_{2\rho}|$$

holds for a sequence $\rho = \rho_k \to 0^+$ with a constant $\gamma > 0$. Indeed, if (3.5) fails, i.e. $u(Y) > \beta_1$, then also $u > \beta_1 > (1-\rho)\beta_1$ on $C_{2\rho}(Y)$ for small $\rho > 0$, and (3.9) also fails.

For $\rho = 1$, (3.9) coincides with the given estimate (3.4). If we show that from (3.9) with some $\rho \in (0,1]$ it follows

$$m\left(\frac{\rho}{2}\right) \le \mu_1 \left(\frac{\rho}{2}\right)^\gamma |C_\rho|, \tag{3.10}$$

then by induction (3.9) holds for $\rho = \rho_k = 2^{-k}$, $k = 0, 1, 2, \cdots$. According to this plan, suppose (3.9) is true for some $\rho \in (0,1]$ and set $h := (1-\rho)\beta_1$. Note that

$$\frac{2}{\rho\beta_1}(u-h) \ge 1 \quad \text{on the set} \quad \left\{u > \left(1 - \frac{\rho}{2}\right)\beta_1\right\}.$$

Using (3.8) and (3.9), we derive

$$\begin{aligned} m\left(\frac{\rho}{2}\right) &= \left|\left\{u > \left(1-\frac{\rho}{2}\right)\beta_1\right\} \cap C_\rho\right| \le \frac{2}{\rho\beta_1}\int_{C_\rho}(u-h)_+\,dX \\ &\le \frac{2N_0}{\rho^2\beta_1}\left|\{u>h\}\cap C_{2\rho}\right|^{1+\alpha} \le \frac{2N_0}{\rho^2\beta_1}\left(\mu_1\rho^\gamma |C_{2\rho}|\right)^{1+\alpha}. \end{aligned}$$

Since $\alpha = \frac{1}{n+2}$ and $|C_\rho| = N\rho^{n+2}$, the last expression will coincide with the right-hand side of (3.10) if we take $\gamma = n+2$ and $\beta_1 = N_1\mu_1^\alpha$ with an appropriate constant $N_1 = N_1(n,\nu) > 0$. By the above arguments

$$u(Y) \le \beta_1 = \beta_1(n,\nu,\mu_1) = \min\left(1, N_1\mu_1^\alpha\right) \to 0 \quad \text{as} \quad \mu_1 \to 0^+.$$

This completes the proof of theorem in the divergence case.

Case (ND). Using rescaling and replacing u by *const* $\cdot\, u$, we may assume without loss of generality that $r = 1$, $M_1(Y) := \sup_{C_1(Y)} u = 1$, and $u(Y) > 0$. Consider the functions

$$v(X) = v(x,t) = u(x,t) + t - s - |x-y|^2 \qquad \text{in} \quad Q := \{v > 0\} \cap C_1(Y).$$

The set Q is nonempty, because $v(Y) = u(Y) > 0$ and $Y \in \partial Q$. From the inequality $v \le u$ in Q and (3.4) it follows

$$|Q| \le |\{u > 0\} \cap C_1(Y)| \le N\mu_1.$$

Since $v \le 0$ on $\partial_p C_1(Y)$, we must have $v = 0$ on $\partial_p Q$. Furthermore, the inequality $Lu \le 0$ yields

$$Lv \le 1 + 2\,\mathrm{tr}\,a \le 1 + 2n\nu^{-1} \quad \text{in} \quad Q.$$

Applying Theorem 2.2 to the function v in Q, we obtain

$$u(Y) = v(Y) \le N\left\|(Lu)_+\right\|_{n+1,Q} \le N|Q|^{\frac{1}{n+1}} \le N_1\mu_1^{\frac{1}{n+1}},$$

with different constants N and N_1, depending only on n and ν. Now the estimate (3.5) follows with

$$\beta_1 = \beta_1(n,\nu,\mu_1) = \min\left(1, N_1\mu_1^\alpha\right), \quad \alpha = \frac{1}{n+1}.$$

Theorem is proved. □

The following theorem follows from Theorem 3.3 by general arguments, which work for both cases (D) and (ND).

THEOREM 3.4. *Let a function* $u \in C^{2,1}(C_r)$, *where* $C_r := C_r(Y)$, $Y = (y,s) \in \mathbb{R}^{n+1}$, $r > 0$, *and let* $Lu \le 0$ *in* C_r. *Then for arbitrary* $p > 0$,

$$u_+^p(Y) \le \frac{N}{|C_r|} \int\limits_{C_r} u_+^p \, dX, \tag{3.11}$$

where $N = N(n,\nu,p) > 0$.

PROOF. Since (3.11) is invariant under rescaling $x \to rx$, $t \to r^2 t$, we may assume $r = 1$. For $X \in C_1 = C_1(Y)$, define

$$d(X) := \sup\{\rho > 0 : \ C_\rho(X) \subset C_1(Y)\}.$$

Obviously, $d(Y) = 1$, so that

$$u(Y) = ud^\gamma(Y) \le M := \sup_{C_1} d^\gamma u, \tag{3.12}$$

where $\gamma := \frac{n+2}{p} > 0$. This choice of γ will be useful below. Note that $d^\gamma u(X)$ is a continuous function on $\overline{C_1}$, which vanishes on $\partial_p C_1$. Therefore,

$$M = d^\gamma u(X_0) \quad \text{for some} \quad X_0 \in \overline{C_1} \setminus \partial_p C_1.$$

It is easy to see that

$$d(X) \ge r_0 := \frac{1}{2} d(X_0) \quad \text{on} \quad C_0 := C_{r_0}(X_0).$$

Hence

$$\sup_{C_0} u \le r_0^{-\gamma} \sup_{C_0} d^\gamma u \le r_0^{-\gamma} M = 2^\gamma u(X_0),$$

and the function $v := u - \frac{1}{2} u(X_0)$ satisfies

$$v(X_0) = 2^{-1} u(X_0) \ge 2^{-\gamma-1} \sup_{C_0} u > 2^{-\gamma-1} \sup_{C_0} v. \tag{3.13}$$

Let $\mu_1 = \mu_1(n,\nu,p) \in (0,1]$ be the constant in Theorem 3.3, which corresponds to $\beta_1 = \beta_1(n,\nu,p) = 2^{-\gamma-1} \in (0,1]$. The previous inequality (3.13) says that v does not belong to $\mathcal{M}(\beta_1, X_0, r_0)$, and therefore by this theorem, the corresponding estimate (3.4) fails, i.e. the set

$$Q_0 := \{v > 0\} \cap C_0 = \left\{u > \frac{1}{2} u(X_0)\right\} \cap C_0$$

has measure $|Q_0| > \mu_1 \cdot |C_0|$. Combining the above estimates, we obtain

$$\begin{aligned} u_+^p(Y) &\le M^p = (d^\gamma u)^p(X_0) = (2r_0)^{\gamma p} u^p(X_0) \\ &\le \frac{(2r_0)^{\gamma p}}{|Q_0|} \int\limits_{Q_0} (2u)^p \, dX \le \frac{2^{\gamma p + p} r_0^{\gamma p}}{\mu_1 |C_0|} \int\limits_{C_1} u_+^p \, dX. \end{aligned}$$

By the choice of γ, we have $r_0^{\gamma p} = r_0^{n+2} = N \cdot |C_0|$, and the estimate (3.11) follows. □

REMARK 3.5. In the above proof, we derived Theorem 3.4 from Theorem 3.3 by general arguments, which do not use further properties of solutions. In turn,

Theorem 3.3 follows easily from Theorem 3.4. Indeed if u satisfies (3.4) and (3.11), then

$$u_+^p(Y) \le \frac{N}{|C_r|} \int\limits_{C_r} u_+^p \, dX \le N \frac{|\{u>0\} \cap C_r|}{|C_r|} \sup_{C_r} u_+^p \le N \mu_1 \sup_{C_r} u_+^p,$$

and $u \in \mathcal{M}(\beta_1, Y, r)$ with $\beta_1 = N(n, \nu, p) \cdot \mu_1^{1/p}$.

4. Second growth theorem

For a fixed point $Y = (y, s) \in \mathbb{R}^{n+1}$ with $s > 0$, and $r > 0$, introduce the *slant cylinder*

$$V_r = V_r(Y) := \left\{ X = (x,t) \in \mathbb{R}^{n+1} : \left| x - \frac{t}{s} y \right| < r,\ 0 < t < s \right\}. \tag{4.1}$$

The following lemma is the main technical tool in this section.

LEMMA 4.1 (Slant Cylinder Lemma). *Let a function $u \in C^{2,1}(\overline{V_r})$ satisfy $Lu \le 0$ in a slant cylinder V_r, which is defined in (4.1) with $Y = (y, s) \in \mathbb{R}^{n+1}$, $s > 0, r > 0$, such that*

$$K^{-1} r |y| \le s \le K r^2, \quad \textit{where} \quad K = const \ge 1. \tag{4.2}$$

In addition, suppose

$$u \le 0 \quad on \quad D_r := B_r(0) \times \{0\}. \tag{4.3}$$

Then

$$u(Y) \le \beta_2 \sup_{V_r(Y)} u_+, \tag{4.4}$$

with a constant $\beta_2 = \beta_2(n, \nu, K) < 1$.

PROOF. Using rescaling, we reduce the proof to the case $r = 1$. Denote $l := s^{-1} y$. By (4.2), we have $K^{-1}|y| \le s \le K$, hence $|l| \le K$. Moreover, approximating u by $u^\varepsilon = G^\varepsilon(u)$ as in the proof of Corollary 2.3, we may assume $u \in C^\infty(\overline{V_1})$ and

$$0 \le u < 1,\ Lu \le 0 \quad \text{on } \overline{V_1}; \quad u \equiv 0 \quad \text{on} \quad D_1 := B_1(0) \times \{0\}. \tag{4.5}$$

Now consider separately cases (D) and (ND).

Case (D). Introduce the function $v = G(u) := -\ln(1-u)$. Obviously, $v \ge 0$ in V, $v \equiv 0$ on D_1. Since

$$G'(u) = (1-u)^{-1} > 0, \quad G''(u) = (1-u)^{-2} > 0 \quad \text{for} \quad u < 1,$$

by Lemma 2.1 we have

$$\begin{aligned} Lu &= G'(u) Lu - G''(u)(aDu, Du) \\ &\le -(1-u)^{-2}(aDu, Du) = -(aDv, Dv) \le -\nu |Dv|^2. \end{aligned}$$

This gives us the estimate

$$a_0 v_t = (D, aDv) + Lv \le (D, aDv) - \nu |Dv|^2 \quad \text{in} \quad V_1. \tag{4.6}$$

Further, take a smooth cut-off function $\eta = \eta(x)$ in $B_1 = B_1(0)$, such that

$$0 \le \eta \le 1 \quad \text{on} \quad B_1 = B_1(0), \quad \eta \equiv 1 \quad \text{on} \quad B_{1/2}, \quad \eta \equiv 0 \quad \text{near} \quad \partial B_1,$$

and define

(4.7)
$$I(t) := \int_{B_1(0)} a_0(x+tl)v(x+tl,t)\,\eta^2(x)\,dx = \int_{B_1(tl)} a_0(x)v(x,t)\,\eta^2(x-tl)\,dx,$$

for $0 \le t \le s$. By the choice of η, the integral functions in (4.7) vanish near ∂B_1, so that we can assume that the integral is taken over $\mathbb{R}^n$. Differentiating with respect to t, we obtain

$$\begin{aligned} I'(t) &= \int [a_0 v_t + (l, Dv)](X_t)\cdot\eta^2(x)\,dx \\ &\le \int \left(|l|\cdot|Dv| + (D, aDv) - \nu|Dv|^2\right)(X_t)\cdot\eta^2(x)\,dx, \end{aligned}$$

where $X_t := (x - tl, t)$. By Schwartz's inequality,

$$|l|\cdot|Dv| \le K|Dv| \le \frac{\nu}{2}|Dv|^2 + \frac{1}{2\nu}K^2.$$

Integrating by parts, we also have

$$\begin{aligned} \int (D, aDv)(X_t)\cdot\eta^2(x)\,dx &= -2\int (D\eta(x), aDv(X_t))\cdot\eta(x)\,dx \\ &\le N\int |Dv|\cdot\eta\,dx \le \frac{\nu}{2}\int |Dv|^2\eta^2\,dx + N_1. \end{aligned}$$

Combining together these estimates, we conclude $I'(t) \le N = N(n,\nu,K)$. In addition, since $v = u = 0$ on $B_1(0)\times\{0\}$, we have $I(0) = 0$. This implies $I(t) \le N_1 t$, and

$$\int_V a_0(x)v(x,t)\eta^2(x-tl)\,dX = \int_0^s I(t)\,dt \le N = N(n,\nu,K).$$

Notice that $a_0 \ge \nu > 0$ and $\eta(x-tl,t) \equiv 1$ on $C_\rho(Y)$ for some small $\rho = \rho(K) > 1$. Hence

$$||v||_{1,C_\rho(Y)} \le N = N(n,\nu,K), \qquad \int_{C_\rho(Y)} v(X)\,dX \le N.$$

By Theorem 3.4 applied to the function v with $p = 1, r = \rho$, we finally obtain $v(Y) \le N_2 = N_2(n,\nu,K)$, and

$$u(Y) = 1 - e^{-v(Y)} \le \beta_2 := 1 - e^{-N_2} < 1.$$

This completes the proof of lemma in the divergence case.

Case (ND). We set

$$v(x,t) := e^{-\gamma t}w^2(x,t), \quad w(x,t) := 1 - |x-tl|^2 > 0 \quad \text{in} \quad V_1,$$

where $\gamma = const > 0$. Then

$$\begin{aligned} Lu &= v_t - (aD, Dv) = e^{-\gamma t}\left[-\gamma w^2 + 2wLw - 2(aDw, Dw)\right] \\ &\le e^{-\gamma t}\left(-\gamma w^2 + Nw - 2\nu|Dw|^2\right) \quad \text{in} \quad V_1. \end{aligned}$$

Notice that $|Dw|^2 = 4|x-tl|^2 = 4(1-w)$. Therefore,

$$Lv \le e^{-\gamma t}\left[-\gamma w^2 + Nw - 8\nu\right] \le 0 \quad \text{in} \quad V_1,$$

provided $\gamma = \gamma(n, \nu, K) > 0$ is chosen large enough. This inequality together with (4.5) imply

$$L(u + v - 1) \le 0 \quad \text{in} \quad V_1.$$

Now compare u and v on the "bottom" $D_1 := B_1(0) \times \{0\}$ and the lateral side $S := \{(x,t) : |x - tl| = 1, \ 0 \le t \le s\}$. From our assumptions it follows

$$u \equiv 0, \quad 0 \le v \le 1 \quad \text{on} \quad D_1; \qquad 0 \le u \le 1, \quad v \equiv 0 \quad \text{on} \quad S.$$

Hence $u+v-1 \le 0$ on the parabolic boundary $\partial_p V = D_1 \cup S$, and by the maximum principle, $u + v - 1 \le 0$ on $\overline{V}_1$. In particular,

$$u(Y) \le 1 - v(Y) = 1 - e^{-\gamma s} \le \beta_2 := 1 - \exp\left(-\gamma K^2\right).$$

Lemma is completely proved. □

THEOREM 4.2 (Second Growth Theorem). *Let a function $u \in C^{2,1}\left(\overline{C_r}\right)$, where $C_r := C_r(Y)$, $Y = (y,s) \in \mathbb{R}^{n+1}$, $r > 0$, and let $Lu \le 0$ in C_r. In addition, suppose*

$$u \le 0 \quad \text{on} \quad D_\rho := B_\rho(z) \times \{\tau\}, \tag{4.8}$$

where

$$B_\rho(z) \subset B_r(y) \quad \text{and} \quad s - r^2 \le \tau \le s - \frac{1}{4}r^2 - \rho^2. \tag{4.9}$$

Then $u \in \mathcal{M}(\beta_2, Y, r)$ with a constant $\beta_2 = \beta_1(n, \nu, \rho/r) < 1$.

PROOF. Using rescaling and parallel translation in $\mathbb{R}^{n+1}$, we reduce the proof to the case $r = 1$ and $(z, \tau) = 0 \in \mathbb{R}^{n+1}$. For an arbitrary $Y' \in C_{1/2}(Y)$, we can apply Lemma 4.1 to the slant cylinder $V_\rho(Y') \subset C_1(Y)$, having in mind that the corresponding constant K in this lemma depends only on ρ. This gives us

$$u(Y') \le \beta_2 \sup_{V_\rho(Y')} u_+ \le \beta_2 \sup_{C_1(Y)} u_+.$$

Since this is true for all $Y' \in C_{1/2}(Y)$, we have

$$\sup_{C_{1/2}(Y)} u_+ \le \beta_2 \sup_{C_1(Y)} u_+,$$

i.e. $u \in \mathcal{M}(\beta_2, Y, 1)$. □

We need one more estimate of same kind, with more explicit dependence of constants on the ratio ρ/r.

LEMMA 4.3. *Let a function $v \in C^{2,1}\left(\overline{C_r}\right)$ satisfy $v \ge 0$, $Lv \ge 0$ in $C_r := C_r(Y)$, $Y = (y,s) \in \mathbb{R}^{n+1}, r > 0$. Then for arbitrary disks $D_\rho := B_\rho(z) \times \{\tau\}$ and $D^0 := B_{r/2}(y) \times \{\sigma\}$, such that $B_\rho(z) \subset B_r(y)$ and*

$$s - r^2 \le \tau < \tau + h^2 r^2 \le \sigma \le s, \ h = const \in (0,1),$$

we have

$$\inf_{D_\rho} v \le \left(\frac{2r}{\rho}\right)^\gamma \inf_{D^0} v, \quad \textit{where} \quad \gamma = \gamma(n, \nu, h) > 0. \tag{4.10}$$

PROOF. Without loss of generality, we assume

$$m := \inf_{D_\rho} v > 0, \quad r = 1, \quad z = 0, \quad \tau = 0, \quad \sigma = h^2,$$

so that $D_\rho = B_\rho(0) \times \{0\}$. Applying the additional linear transformation along the t-axis, we can also reduce the proof to the case $h = 1$. After these preparations, fix the integer $k \geq 0$ in such a way that $2^{-k-1} < \rho \leq 2^{-k}$, and introduce the disks

$$D^j := B_{2^{-j-1}}\left(2^{-j} y\right) \times \left\{4^{-j}\right\} \quad \text{for} \quad j = 0, 1, \cdots, k.$$

We can apply Theorem 4.2 to the function $u := 1 - m^{-1} v$ in $C_{2^{-k-1}}(Y')$, with an arbitrary $Y' \in D^k$, and $K = 3$. This gives us

$$\sup_{D^k} u \leq \beta_2 = \beta_2(n, \nu) < 1,$$

or, equivalently,

$$\inf_{D_\rho} v = m \leq (1 - \beta_2)^{-1} \inf_{D^k} v = 2^\gamma \inf_{D^k} v,$$

where $\gamma = \gamma(n, \nu) := -\log_2(1 - \beta_2) > 0$. Similarly, if $k \geq 1$, we also have

$$\inf_{D^j} v \leq 2^\gamma \inf_{D^{j-1}} v \quad \text{for} \quad j = 1, 2, \cdots, k.$$

Combining these inequalities, we obtain

$$\inf_{D_\rho} v \leq 2^\gamma \inf_{D^k} v \leq 2^{2\gamma} \inf_{D^{k-1}} v \leq \cdots \leq 2^{(k+1)\gamma} \inf_{D^0} v \leq \left(\frac{2}{\rho}\right)^\gamma \inf_{D^0} v.$$

□

5. Third growth theorem

The first growth theorem, Theorem 3.1, describes the relation between the constants β_1 and μ_1 in (3.1) and (3.4) in such a way that $\beta_1 = \beta_1(n, \nu, \mu_1) \to 0^+$ as $\mu_1 \to 0^+$. In this section we show that, roughly speaking, $\beta_1 < 1$ if $\mu_1 < 1$. The proof of this fact, which we call the *third growth theorem*, is based on the first and second growth theorems. We also need two covering lemmas, which are similar to those in [**KS**] and [**S80**].

LEMMA 5.1. *Let a constant $\mu_0 \in (0, 1)$ be fixed. For an arbitrary measurable set $\Gamma \subset \mathbb{R}^{n+1}$ with finite measure $|\Gamma|$, introduce the family of cylinders*

(5.1) $$\mathcal{A} := \{C = C_r(Y) : |C \cap \Gamma| \geq (1 - \mu_0) \cdot |C|\}.$$

Then the open set $E := \cup\{C : C \in \mathcal{A}\}$ satisfies

(5.2) $$|\Gamma \setminus E| = 0 \quad \text{and} \quad |E| \geq q_0 |\Gamma|, \quad \text{where} \quad q_0 := 1 + 3^{-n-1} \mu_0 > 1.$$

PROOF. The first statement $|\Gamma \setminus E| = 0$ follows from the fact that almost every point of Γ is a point of density. Indeed, suppose $|\Gamma \setminus E| > 0$. Then one can choose a cylinder $C^* := B_{1/m}(y) \times \left(s - \frac{1}{m}, s\right)$, with natural m, such that

(5.3) $$|C^* \cap (\Gamma \setminus E)| > (1 - \mu_0) \cdot |C^*|.$$

Since C^* is a union of m disjoint *parabolic* cylinders

$$C_k^* := B_{1/m}(y) \times \left(s - \frac{k+1}{m^2}, s - \frac{k}{m^2}\right), \quad k = 0, 1, \cdots, m - 1,$$

the inequality (5.3) must be true for some C_k^* instead of C^*. But then $C_k^* \in \mathcal{A}$, $C_k^* \subset E$, $C_k^* \cap (\Gamma \setminus E)$ is empty and (5.3) cannot be true for C_k^*. This contradiction proves $|\Gamma \setminus E| = 0$.

Further, for each $C = C_r(Y) \in \mathcal{A}$ with $|C \cap \Gamma| > (1-\mu_0) \cdot |C|$, we can continuously increase r until we get the equality $|C \cap \Gamma| = (1-\mu_0) \cdot |C|$ (note that $|\Gamma| < \infty$). Therefore, we can write

$$(5.4) \quad E = \cup \{C : C \in \mathcal{A}_0\}, \quad \mathcal{A}_0 := \{C = C_r(Y) : |C \cap \Gamma| = (1-\mu_0) \cdot |C|\}.$$

Next, we reproduce the well known argument in the proof of the classical Vitali covering theorem (with parabolic cylinders instead of usual balls or cubes). We construct a (countable or finite) sequence of cylinders $C^k, k = 1, 2, \cdots$, as follows. For this, we denote $R_1 := \sup \{r : C_r(Y) \in \mathcal{A}_0\}$. Easy compactness argument shows that this supreme is attained for some cylinder $C_{R_1}(Y_1) \in \mathcal{A}_0$. We pick an arbitrary one of (possibly many) such cylinders: $C^1 := C_{R_1}(Y_1)$. Assuming that $C^i := C_{R_i}(Y_i)$ for $i = 1, 2, \cdots, k$ have been already selected for some $k \geq 1$, we set

$$\mathcal{A}_{k+1} := \left\{C = C_r(Y) \in \mathcal{A}_0 : C \cap C^i = \emptyset, \ i = 1, 2, \cdots, k\right\}.$$

If $\mathcal{A}_{k+1}$ is nonempty, then we denote

$$R_{k+1} := \sup \{r > 0 : C_r(Y) \in \mathcal{A}_{k+1} \text{ for some } Y\}$$

this supreme is attained for $C^{k+1} := C_{R_{k+1}}(Y_{k+1}) \in \mathcal{A}_{k+1}$.

In the case when all the sets $\mathcal{A}_0 \supset \mathcal{A}_1 \supset \mathcal{A}_2 \supset \cdots$ are nonempty, we obtain a countable sequence of cylinders $C^i = C_{R_i}(Y_i), i = 1, 2, \cdots$. If however, $\mathcal{A}_i \neq \emptyset$ for $i = 1, 2, \cdots, k$ and $\mathcal{A}_{k+1} = \emptyset$, then we have a finite sequence of cylinders $C^i, i = 1, 2, \cdots, k$. In the latter case we set, by definition, $R_{k+1} = R_{k+2} = \cdots = 0$. We note that by construction the cylinders C^i are pairwise disjoint, $R_1 \geq R_2 \geq \cdots$, and $R_i \to 0$ as $i \to \infty$.

Take an arbitrary cylinder $C_r(Y) \in \mathcal{A}_0$. We have $R_1 \geq R_2 \geq \cdots \geq R_k \geq r > R_{k+1}$ for some integer $k \geq 1$. Since $r > R^{k+1}$, the cylinder $C_r(Y)$ does not belong to $\mathcal{A}_{k+1}$, and therefore $C_r(Y) \cap C^i \neq \emptyset$ for some $i \leq k$. Since $r < R_i$, this implies $C_r(Y) \subset \tilde{C}^i$, where

$$\tilde{C} := B_{3\rho}(z) \times \left(s - 2\rho^2, s + \rho^2\right) \quad \text{for} \quad C = C_\rho(z, \tau) = B_\rho(z) \times \left(s - \rho^2, s\right).$$

Thus arbitrary $C_r(Y) \in \mathcal{A}_0$ is a subset of $\tilde{C}^i$ for some $i \geq 1$. From (5.4) we conclude

$$E \subset \bigcup_i \tilde{C}^i, \quad |E| \leq \sum_i \left|\tilde{C}^i\right| = 3^{n+1} \sum_i \left|C^i\right|.$$

On the other hand, since $C^i \in \mathcal{A}_0$ are pairwise disjoint,

$$|E \setminus \Gamma| \geq \sum_i \left|(E \setminus \Gamma) \cap C^i\right| = (1-\mu_0) \sum_i \left|C^i\right|.$$

From the previous relations we obtain

$$\frac{|E|}{|\Gamma|} = 1 + \frac{|E \setminus \Gamma|}{|\Gamma|} \geq 1 + \frac{|E \setminus \Gamma|}{|E|} \geq 1 + \frac{\mu_0}{3^{n+1}} = q_0 > 1.$$

The lemma is proved. □

LEMMA 5.2. *For a fixed constant* $K_1 > 1$ *and any standard cylinder* $C = C_r(Y) = B_r(y) \times (s - r^2, s)$ *denote* $\hat{C} := B_r(y) \times (s + r^2, s + K_1 r^2)$. *Then for arbitrary family* $\mathcal{A}$ *of standard cylinders* $C = C_r(Y)$, *the* $n+1$*-dimensional measures of the sets* $E := \cup \{C : C \in \mathcal{A}\}$ *and* $\hat{E} := \cup \{\hat{C} : C \in \mathcal{A}\}$ *satisfy*

$$\left|\hat{E}\right| \geq q_1 |E|, \quad \text{where} \quad q_1 := \frac{K_1 - 1}{K_1 + 1}. \tag{5.5}$$

PROOF. By Fubini's theorem,

$$|E| = \int |E_x| \, dx, \quad \left|\hat{E}\right| = \int \left|\hat{E}_x\right| dx,$$

where

$$E_x := \left\{t \in \mathbb{R}^1 : (x, t) \in E\right\}, \quad \hat{E}_x := \left\{t \in \mathbb{R}^1 : (x, t) \in \hat{E}\right\}, \quad x \in \mathbb{R}^n.$$

Therefore, for the proof of lemma it suffices to get the estimate $\left|\hat{E}_x\right| \geq q_1 |E_x|$ for all $x \in \mathbb{R}^n$.

Fix an arbitrary x for which E_x is nonempty. For this x, the open set $\hat{E}_x$ is a union of disjoint open intervals $\hat{I}_k$, $k = 1, 2, \cdots$. If $t \in E_x$, then (x, t) belongs to a cylinder $C = B_r(y) \times (s - r^2, s)$ in $\mathcal{A}$, and $(s + r^2, s + K_1 r^2) \subset \hat{I}_k$ for some k. Obviously, $r \leq r_k := (K_0 - 1)^{-1} \left|\hat{I}_k\right|$. We also choose s_k such that $\hat{I}_k = (s_k + r_k^2, s_k + K_1 r_k^2)$. Then

$$t \in (s - r^2, s) \subset J_k := (s_k - r_k^2, s_k + K_1 r_k^2).$$

Our argument shows that $E_x \subset \bigcup_k J_k$, hence

$$q_1 |E_x| \leq q_1 \left|\bigcup_k J_k\right| \leq q_1 \sum_k |J_k| = \sum_k \left|\hat{I}_k\right| = \left|\hat{E}_x\right|.$$

This completes the proof of lemma. □

THEOREM 5.3 (Third Growth Theorem). *Let a function* $u \in C^{2,1}(\overline{C_r})$, *where* $C_r := C_r(Y)$, $Y = (y, s) \in \mathbb{R}^{n+1}$, $r > 0$, *and let* $Lu \leq 0$ *in* C_r. *In addition, suppose*

$$\left|\{u > 0\} \cap C^0\right| \leq \mu \cdot \left|C^0\right| \tag{5.6}$$

where

$$C^0 := C_{r/2}(Y^0), \quad Y^0 := \left(y, s - \frac{3}{4} r^2\right), \quad \mu = const < 1.$$

Then $u \in \mathcal{M}(\beta, Y, r)$ *with a constant* $\beta = \beta(n, \nu, \mu) < 1$.

PROOF. For certainty, we assume $Y = (y, s) = 0$ and $r = 2$. As before, the general case is reduced to this one by a linear transformation. Note that $C^0 := B_1(0) \times (-4, -3)$, hence $\left|C^0\right| = |B_1|$ depends only on n.

Consider the set $\Gamma := \{u \leq 0\} \cap C^0$. From (5.6) it follows

$$|\Gamma| \geq (1 - \mu) \left|C^0\right| = (1 - \mu) |B_1| =: c_0 = c_0(n, \mu) > 0. \tag{5.7}$$

In Theorem 3.3, the constant $\beta_1 = \beta_1(n, \nu, \mu) \to 0^+$ as $\mu \to 0^+$. Fix a constant $\mu_0 = \mu_0(n, v) \in (0, 1)$ such that the corresponding value $\beta_1(n, \nu, \mu_0) \leq \frac{1}{2}$. For this constant μ_0 and the given set Γ, consider the family of cylinders $\mathcal{A}$ defined by the formula (5.1). By Lemma 5.1, the set $E := \cup \{C : C \in A\}$ has measure

$$|E| \geq q_0 |\Gamma|, \quad \text{where} \quad q_0 := 1 + 3^{-n-1}\mu_0 > 1.$$

Denote $\varepsilon_0 := 3^{-n-1}\mu_0$, so that $q_0 = 1 + 3\varepsilon_0$. Next, choose the constant $K_1 > 0$ satisfying the equality

$$q_1 = \frac{K_1 - 1}{K_1 + 1} = \frac{1 + 2\varepsilon_0}{1 + 3\varepsilon_0}, \quad \text{i.e.} \quad K_1 = K_1(n, \nu) := 5 + 2\varepsilon_0^{-1}.$$

Combining Lemmas 5.1 and 5.2, we find that the corresponding set $\hat{E}$ has measure

$$\left|\hat{E}\right| \geq q_1 |E| \geq q_0 q_1 |\Gamma| = (1 + 2\varepsilon_0) |\Gamma|. \tag{5.8}$$

Further, choose another open cylinder C^1 such that

$$C^0 \subset C^1, \quad \left|C^1 \setminus C^0\right| \leq \varepsilon_0 c_0, \quad \text{and} \quad dist\left(C^0, \partial C^1\right) \geq c_1 = c_1(n, \nu, \mu) > 0. \tag{5.9}$$

Consider separately cases (a) $\hat{E} \setminus C^1 \neq \emptyset$ and (b) $\hat{E} \setminus C^1 = \emptyset$.

(a) $\hat{E} \setminus C^1 \neq \emptyset$. This can be true only if some cylinders $C \in \mathcal{A}$ are large enough: there exists $C = C_r(Y) \in \mathcal{A}$ with $r \geq r_0 = r_0(n, \nu, \mu) > 0$. Note that

$$|\{u > 0\} \cap C| \leq |C \setminus \Gamma| \leq \mu_0 |C| \quad \text{for} \quad C \in \mathcal{A}.$$

By Theorem 3.3 and the choice of μ_0,

$$\sup_{C_{\frac{r}{2}}(Y)} u \leq \frac{1}{2} \sup_{C_r(Y)} u_+ \leq \frac{1}{2} M \quad \text{for} \quad C = C_r(Y) \in \mathcal{A}, \tag{5.10}$$

where $M := \sup_{C_2} u_+$, hence

$$u - \frac{1}{2} M \leq 0 \quad \text{on} \quad D := B_{r/2}(y) \times \{s\} \quad \text{for} \quad C_r(Y) \in \mathcal{A},\ Y = (y, s).$$

In our case, we can apply Theorem 4.2 to the function $u - \frac{1}{2}M$ with $\rho = \frac{1}{2}r_0$. By this theorem, $u - \frac{1}{2}M \in \mathcal{M}(\beta_2, 0, 2)$ with $\beta_2 < 1$ depending only on n, ν, μ. Then

$$\begin{aligned} \sup_{C_1} u &= \frac{1}{2}M + \sup_{C_1}(u - \frac{1}{2}M) \\ &\leq \frac{1}{2}M + \beta_2 \sup_{C_2}(u - \frac{1}{2}M)_+ = \frac{1}{2}(1 + \beta_2) M, \end{aligned}$$

and $u \in \mathcal{M}(\beta_0, 0, 2)$ with $\beta_3 := \frac{1}{2}(1 + \beta_2) < 1$.

(b) $\hat{E} \setminus C^1 = \emptyset$. In this case $\hat{E} \subset C^1$, and by the estimates (5.7)–(5.9), the set $\Gamma_1 := \hat{E} \cap C^0$ has measure

$$|\Gamma_1| = \left|\hat{E} \cap C^0\right| \geq \left|\hat{E}\right| - \left|C^1 \setminus C^0\right| \geq (1 + 2\varepsilon_0) |\Gamma| - \varepsilon_0 c_0 \geq (1 + \varepsilon_0) |\Gamma|. \tag{5.11}$$

The previous argument in (a) shows that for arbitrary $C \in \mathcal{A}$, from (5.10) it follows $u \leq \beta_0 M$ on $\hat{C}$ with $\beta_0 = \beta_0(n, \nu, \mu) < 1$. Since such sets $\hat{C}$ cover Γ_1, we have $u \leq \beta_3 M$ on Γ_1. Combining this estimate with (5.11), we obtain

$$\left|\{u \leq \beta_0 M\} \cap C^0\right| \geq (1 + \varepsilon_0) \left|\{u \leq 0\} \cap C^0\right|. \tag{5.12}$$

We have proved that either (a) $u \leq \beta_0 M$ on C_1, or (b) u satisfies (5.12). Starting from $u_0 := u$, $M_0 := M$, define

$$u_{k+1} := u_k - \beta_0 M_k, \quad M_{k+1} := \sup_{C_2} u_{k+1} \quad \text{for} \quad k = 0, 1, 2, \cdots.$$

It is easy to see that

$$u_k = u - \left[1 - (1-\beta_0)^k\right] M, \quad M_k = (1-\beta_0)^k M \qquad \text{for all} \quad k.$$

If the alternative (b) holds for all u_k with $k = 0, 1, 2, \cdots, m$, then for such k

$$\begin{aligned} \left|C^0\right| &\geq \left|\{u_m \leq 0\} \cap C^0\right| = \left|\{u_{m-1} \leq \beta_0 M_{m-1}\}\right| \geq (1+\varepsilon_0) \left|\{u_{m-1} \leq 0\} \cap C^0\right| \\ &\geq \cdots \geq (1+\varepsilon_0)^m \left|\{u_0 \leq 0\} \cap C^0\right| \geq (1+\varepsilon_0)^m c_0. \end{aligned}$$

If we take m such that $(1+\varepsilon_0)^m c_0 > \left|C^0\right|$, then (b) fails for some u_k with $k \leq m$. Then u_k satisfies (a) $u_k \leq \beta_0 M_k$ on C_1. In other words, $u_{k+1} \leq 0$ on C_1, and hence

$$u = u_{k+1} + \left[1 - (1-\beta_0)^{k+1}\right] M \leq \left[1 - (1-\beta_0)^{m+1}\right] M \quad \text{on} \quad C_1.$$

The last inequality exactly means

$$u \in \mathcal{A}(\beta, 0, 2) \quad \text{with} \quad \beta := 1 - (1-\beta_0)^{m+1} < 1.$$

□

COROLLARY 5.4. *Let a function $v \in C^{2,1}\left(\overline{C_r}\right)$ satisfy $v \geq 0$, $Lv \geq 0$ in C_r, and in addition, $\left|\{v \geq 1\} \cap C^0\right| \geq (1-\mu) \cdot \left|C^0\right|$. Then $v \geq 1 - \beta > 0$ on $C_{r/2}$. Here $\beta = \beta(n, \nu, \mu) < 1$ for $\mu < 1$.*

PROOF. We assume $r = 2$. The function $u := 1 - v$ satisfies $Lv \leq 0$ in C_2, and

$$\begin{aligned} \left|\{u > 0\} \cap C^0\right| &= \left|\{v < 1\} \cap C^0\right| = |C_0| - \left|\{v \geq 1\} \cap C^0\right| \\ &\leq |C_0| - (1-\mu) \cdot \left|C^0\right| = \mu \cdot \left|C^0\right|. \end{aligned}$$

By the previous theorem,

$$\sup_{C_1} u \leq \beta \sup_{C_2} u_+ \leq \beta, \quad \text{and} \quad \inf_{C_1} v = 1 - \sup_{C_1} u \geq 1 - \beta.$$

□

6. Proof of Theorem 1.5. Some generalizations and applications

PROOF. We may assume $Y = 0, r = 1$, so that $C_1 = B_1(0) \times (-1, 0)$, $C^0 = B_1(0) \times (-3, -2)$. Obviously,

$$d(X) := \sup\{\rho > 0 : \ C_\rho(X) \subset C_2(0)\} \geq 1 \quad \text{on} \quad C^0.$$

Therefore,

$$\sup_{C^0} u \leq M := \sup_{C^1} d^\gamma u, \quad \text{where} \quad C^1 := B_2(0) \times (-3, -2),$$

and $\gamma > 0$ is the constant in Lemma 4.3 corresponding to $h = \frac{1}{2}$. Following the proof of Theorem 3.4, we have

$$M = d^\gamma u(X_0) \quad \text{for some} \quad X_0 = (x_0, t_0) \in \overline{C^1} \setminus \partial_p C_1,$$

and then for

$$\rho := \frac{1}{4} d(X_0) \in \left(0, \frac{1}{2}\right], \quad C_0 := C_\rho(X_0), \quad Q_0 := \left\{u > \frac{1}{2} u(X_0)\right\} \cap C_0,$$

we have $|Q_0| > \mu_1 |C_0|$ with a constant $\mu_1 = \mu_1(n, \nu) > 0$. Now we can apply Corollary 5.4 with

$$v = \frac{2}{u(X_0)} u, \quad C_r = C_{2\rho}(Y_0), \quad Y_0 := \left(x_0, t_0 + 2\rho^2\right), \quad C^0 = C_0, \quad 1 - \mu = \mu_1.$$

As a result, we get the estimate

$$u \geq \beta u(X_0) \quad \text{on} \quad C_\rho(Y_0) \quad \text{with} \quad \beta = \beta(n,\nu) > 0.$$

Further, we use Lemma 4.3 with

$$v = u, \quad r = 2, \quad D_\rho = B_\rho(x_0) \times \{t_0 + \rho^2\} \subset \overline{C_\rho(Y_0)},$$

and $D^0 = B_1(0) \times \{\tau\}$ with an arbitrary $\tau \in (-1,0)$. This yields

$$\beta u(X_0) \leq \inf_{D_\rho} u \leq \left(\frac{4}{\rho}\right)^\gamma \inf_{C_1(0)} u.$$

Finally,

$$\sup_{C^0} u \leq M = d^\gamma u(X_0) = (4\rho)^\gamma u(X_0) \leq \beta^{-1} 4^{2\gamma} \inf_{C_1(0)} u.$$

This prove the theorem with $N = N(n,\nu) = \beta^{-1}4^{2\gamma}$. □

REMARK 6.1. The main statements, such as the Harnack inequality and growth theorems, are naturally extended to more general classes, which are different the cases (D) and (ND). These are classes obtained by closure of the set $C^{2,1}$ (or C^∞) with respect to the integral norms discussed in Section 2. We demonstrate this approach in the case (ND) for the proof of the Harnack inequality. We assume

$$u \geq 0, \quad Lu := u_t - \sum_{i,j} a_{ij} D_{ij} u = 0 \quad \text{in} \quad C_{2r},$$

where the coefficients a_{ij} are measurable functions, and u belongs to the Sobolev space $W^{2,1}_{n+1}(C_{2r}) \cap C\left(\overline{C_{2r}}\right)$, i.e. $u_t, D_{ij}u \in L^{n+1}$. One can approximate a_{ij} and u by smooth functions $a^\varepsilon_{ij} \to a_{ij}$ (a. e.) and $u^\varepsilon \to u$ (in $W^{2,1}_{n+1}$) as $\varepsilon \to 0^+$. Then

$$f^\varepsilon := L^\varepsilon u^\varepsilon := u^\varepsilon_t - \sum_{i,j} a^\varepsilon_{ij} D_{ij} u^\varepsilon \to 0 \quad (\text{in } L^{n+1}) \text{ as } \varepsilon \to 0^+.$$

Relying on the existence results for equations with smooth coefficients, we can write $u^\varepsilon = v^\varepsilon + w^\varepsilon$, where

$$\begin{array}{rclll} L^\varepsilon v^\varepsilon & = & 0 \ \text{ in } \ C_{2r}, & v^\varepsilon = u^\varepsilon & \text{on } \ \partial_p C_{2r}; \\ L^\varepsilon v^\varepsilon & = & f^\varepsilon \ \text{ in } \ C_{2r}, & v^\varepsilon = 0 & \text{on } \ \partial_p C_{2r}. \end{array}$$

By Theorem 2.4, $v^\varepsilon \to 0$ in L^∞, while v^ε satisfies the Harnack inequality (1.2). By easy limit passage, (1.2) also holds for the original function u.

REMARK 6.2. In the divergence case, we do not assume that $a_{ij} = a_{ji}$. This makes it easy to extend the results to more general equations

$$L'u = a_0 u_t - \sum_{i,j=1}^n D_i(a_{ij} D_j u) - \sum_{j=1}^n b_j D_j u - \sum_{i=1}^n D_i(c_i u) = 0, \tag{6.1}$$

which include the lower order terms with bounded coefficients b_j, c_i. We set

$$a^0_{00} = A = const > 0, \quad a^0_{i0} = c_i, \quad a^0_{0j} = b_j, \quad a^0_{ij} = a_{ij}, \quad \text{for} \quad i,j = 1, \cdots, n.$$

Then the $(n+1) \times (n+1)$-matrix $\left[a^0_{ij}\right]^n_{i,j=0}$ satisfies the condition the uniform parabolicity condition (U) with a smaller constant $\nu > 0$, provided $A > 0$ is large enough. Introduce a new variable $x_0 \in \mathbb{R}^1$. Then the function

$$u^0(x_0, x, t) = e^{x_0 + At} u(x,t),$$

satisfy the parabolic equation without lower order terms:

$$L^0u^0 = a_0u_t^0 - \sum_{i,j=0}^{n} D_i(a_{ij}^0 D_j u^0) = Au^0 + e^{x_0+At}L'u - Au^0 = 0.$$

Thus the Harnack inequality and some other results are easily extended to more general equations (6.1).

References

[A] A. D. Aleksandrov, *Uniqueness conditions and estimates for the solutions of the Dirichlet problem*, Vestnik Leningrad. Univ. **18**, no. 3 (1963), 5–29 in Russian; English translation in Amer. Math. Soc. Trans. **68**, no. 2 (1968), 89–119.

[A68] D. G. Aronson, *Non-negative solutions of linear parabolic equations*, Ann. Scuola Norm. Sup. Pisa **22** (1968), 607–694.

[AS] D. G. Aronson and J. Serrin, *Local behavior of solutions of quasilinear parabolic equations*, Arch. Rational Mech. Anal. **25** (1967), 81–122.

[DG] E. De Giorgi, *Sulla differenziabilità e l'analiticità delle estremali degli integrali multipli regolari*, Mem. Accad. Sci. Torino Cl. Sci. Fis. Mat. Natur. (3) **3** (1957), 25–43.

[F] A. Friedman, *Partial Differential Equations of Parabolic Type*, Prentice Hall, 1964.

[FS] E. B. Fabes and M. V. Safonov, *Behavior near the boundary of positive solutions of second order parabolic equations*, J. Fourier Anal. and Appl., Special Issue: Proceedings of the Conference El Escorial 96 (1998), 871–882.

[FSY] E. B. Fabes, M. V. Safonov and Yu Yuan, *Behavior near the boundary of positive solutions of second order parabolic equations. II*, Trans. Amer. Math. Soc. **351**, no. 12 (1999), 4947–4961.

[FSt] E. B. Fabes and D. W. Stroock, *A new proof of Moser's parabolic Harnack inequality using the old idea of Nash,* Arch. Rational Mech. Anal. **96**, no. 4 (1986), 327–338.

[GT] D. Gilbarg and N. S. Trudinger, *Elliptic Partial Differential Equations of second Order*, 2nd ed., Springer-Verlag, Berlin–Heidelberg–New York–Tokyo, 1983.

[I] A. V. Ivanov, *The Harnack inequality for generalized solutions of second-order quasilinear parabolic equations,* Trudy Mat. Inst. Steklov **102** (1967), 51–84.

[K76] N. V. Krylov, *Sequences of convex functions and estimates of the maximum of the solution of a parabolic equation,* Sibirsk. Math. Zh. **17** (1976), 290–303 in Russian; English translation in Siberian Math. J. **17** (1976), 226–236.

[K] N. V. Krylov, *Nonlinear Elliptic and Parabolic Equations of Second Order*, Nauka, Moscow, 1985 in Russian; English translation: Reidel, Dordrecht, 1987.

[KS] N. V. Krylov and M. V. Safonov, *A certain property of solutions of parabolic equations with measurable coefficients*, Izvestia Akad. Nauk SSSR, ser. Matem. **44**, No. 1 (1980), 161–175 in Russian; English translation in Math. USSR Izvestija, **16**, no. 1 (1981), 151–164.

[LE] E. M. Landis, *Second Order Equations of Elliptic and Parabolic Type,* "Nauka", Moscow, 1971 in Russian; English transl.: Amer. Math. Soc., Providence, RI, 1997.

[L] G. M. Lieberman, *Second Order Parabolic Differential Equations*, World Scientific, 1996.

[LSU] O. A. Ladyzhenskaya, V. A. Solonnikov and N. N. Ural'tseva, *Linear and Quasi-linear Equations of Parabolic Type*, Nauka, Moscow, 1967 in Russian; English transl.: Amer. Math. Soc., Providence, RI, 1968.

[LSW] W. Littman, G. Stampacchia and H. W. Weinberger, *Regular points for elliptic equations with discontinuous coefficients,* Ann. Scuola Norm. Sup. Pisa (3) **17** (1963), 43–77.

[M61] J. Moser, *On Harnack's theorem for elliptic differential equation,* Comm. Pure Appl. Math. **14** (1961), 577-591.

[M64] J. Moser, *A Harnack inequality for parabolic differential equations*, Comm. Pure and Appl. Math. **17** (1964), 101–134; and correction in: Comm. Pure and Appl. Math. **20** (1967), 231–236.

[N] J. Nash, *Continuity of solutions of parabolic and elliptic equations,* Amer. J. Math. **80** (1958), 931–954.

[PE] F. O. Porper and S. D. Eidelman, *Two-sided estimates of fundamental solutions of second-order parabolic equations, and some applications*, Uspekhi Mat. Nauk **39**, no. 3 (1984), 107–156 in Russian; English transl. in Russian Math. Surveys **39**, no. 3 (1984), 119–178.

[S80] M. V. Safonov, *Harnack inequality for elliptic equations and the Hölder property of their solutions*, Zap. Nauchn. Sem. Leningrad. Otdel. Mat. Inst. Steklov (LOMI) **96** (1980), 272–287 in Russian; English translation in J. Soviet Math. **21**, no. 5 (1983), 851–863.

[S98] M. V. Safonov, *Estimates near the boundary for solutions to second order parabolic equaitons,* Invited Lecture at the ICM in Berlin, August 18–27, 1998. In: Documenta Matematica, Extra Volume ICM 1998, vol. 1, pp. 637–647.

[SY] M. V. Safonov and Yu Yuan, *Doubling properties for second order parabolic equations*, Annals of Mathematics **150** (1999), 313–327.

[SJ] J. Serrin, *Local behavior of solutions of quasi-linear equations,* **111** (1964), 247–302.

[T] N. S. Trudinger, *Pointwise estimates and quasilinear parabolic equations*, Comm. Pure and Appl. Math. **21** (1968), 205–226.

[TK] K. Tso, *On an Aleksandrov-Bakelman maximum principle for second-order parabolic equations,* Comm. Partial Differential Equations **10** (1985), 543–553.

School of Mathematics, University of Minnesota, Minneapolis, MN 55455
E-mail address: `ferretti@math.umn.edu`

School of Mathematics, University of Minnesota, Minneapolis, MN 55455
E-mail address: `safonov@math.umn.edu`

Contemporary Mathematics
Volume **277**, 2001

Absolute Continuity of Generalized Periodic Schrödinger Operators

Zhongwei Shen

ABSTRACT. We study the spectrum of the generalized Schrödinger operator $-\Delta + \mu$ in $\mathbb{R}^d$, $d \geq 2$ where μ is a signed periodic Radon measure. Under the condition that $|\mu|(B(\mathbf{x}, r)) \leq C\, r^{d-2+\varepsilon}$ for all $\mathbf{x} \in \mathbb{R}^d$, $0 < r < 1$ where $\varepsilon > 1$ if $d \geq 3$, and $\varepsilon > 0$ if $d = 2$, we show that the spectrum of $-\Delta + \mu$ is purely absolutely continuous. We also consider the elliptic operator $\text{div}(\omega(\mathbf{x})\nabla)$ in $\mathbb{R}^d$ with a positive periodic scalar function $\omega(\mathbf{x})$. We prove that, if $\omega \in C^1(\mathbb{R}^d)$, then the spectrum of $\text{div}(\omega(\mathbf{x})\nabla)$ is purely absolutely continuous.

1. Introduction

Consider the generalized Schrödinger operator

$$(1.1) \qquad -\Delta + \mu \qquad \text{in} \quad \mathbb{R}^d, \quad d \geq 2$$

with a signed Radon measure μ. We assume that μ is periodic, i.e., there exists a basis $\{\mathbf{e}_j\}_{j=1}^d$ of $\mathbb{R}^d$ such that

$$(1.2) \qquad \mu(S + \mathbf{e}_j) = \mu(S), \qquad j = 1, 2, \ldots, d$$

for any Borel measurable subset S of $\mathbb{R}^d$. Throughout this paper we will also impose the following regularity condition on the total variation measure $|\mu|$ of μ,

$$(1.3) \qquad |\mu|(B(\mathbf{x}, r)) \leq C\, r^{d-2+\varepsilon} \qquad \text{for all } \mathbf{x} \in \mathbb{R}^d, \quad 0 < r \leq 1,$$

and some $\varepsilon > 0$, where $B(\mathbf{x}, r)$ denotes the ball centered at $\mathbf{x}$ with radius r. Under conditions (1.2) and (1.3), it will be shown in Lemma 2.4 that

$$(1.4) \qquad \int_{\mathbb{R}^d} |\psi|^2 \, d|\mu| \leq \alpha \int_{\mathbb{R}^d} |\nabla \psi|^2 \, d\mathbf{x} + C_\alpha \int_{\mathbb{R}^d} |\psi|^2 \, d\mathbf{x}$$

for any $\alpha > 0$ and $\psi \in C^1(\mathbb{R}^d)$. Let $\mathbf{D} = -i\nabla$ and $A = (a_{jl})_{d \times d}$ be a symmetric, positive definite matrix with real constant entries. It follows from (1.4) that the quadratic form associated with $\sum_{j,l} D_j a_{jl} D_l + \mu$ defines a unique self-adjoint operator on $L^2(\mathbb{R}^d)$, which we denote by $\mathbf{D}A\mathbf{D}^T + \mu$. The following is the main result of the paper.

1991 *Mathematics Subject Classification.* Primary 35J10.

The author was supported in part by NSF Grant DMS-9732894.

Main Theorem. *Let $A = (a_{jl})_{d\times d}$ be a symmetric, positive definite matrix with real constant entries. Let μ be a signed periodic Radon measure on $\mathbb{R}^d$ satisfying (1.3) for some $\varepsilon > 0$ if $d = 2$, and some $\varepsilon > 1$ if $d \geq 3$. Then the spectrum of $\mathbf{D}A\mathbf{D}^T + \mu$ is purely absolutely continuous.*

For the Schrödinger operator $-\Delta + V$ with a real periodic function potential V, the absolute continuity of its spectrum was first established by L. Thomas [**21**] for $d = 3$ and $V \in L^2_{\text{loc}}$. Thomas' result was subsequently extended to $\mathbb{R}^d$, $d \geq 2$ by M. Reed and B. Simon [**14, Vol.IV**]. Recent work on the problem of absolute continuity for the Dirac operator and the magnetic Schrödinger operator with periodic potentials may be found in [**1-7,12-13,19**]. In [**16**] the author obtained the absolute continuity of $-\Delta + V$ under the condition $d \geq 3$, $V \in L^{d/2}_{\text{loc}}(\mathbb{R}^d)$. This is best possible in the context of L^p spaces. We also gave the optimal condition for the absolute continuity of $-\Delta + V$ with a periodic potential in Lorentz spaces in [**16**]. The results in [**16**] were further extended to the Schrödinger operator $-\Delta + V$ with potential V in the Morrey-Campanato class [**17**] and in the Kato class for $d = 2$ or 3 [**18**].

We point out that, if $d\mu = V(\mathbf{x})\,d\mathbf{x}$, then μ satisfies the assumption of the main theorem if $V \in L^p_{\text{loc}}(\mathbb{R}^d)$ where $p > 1$ for $d = 2$, and $p > d$ for $d \geq 3$. Thus the main theorem is weaker than the known results in [**3,14,16,18,21**]. However our main theorem in this paper allows μ to be a (singular) measure supported on a set of Lebesgue measure zero in $\mathbb{R}^d$. In particular, if $d = 2$, Γ is the union of a finite number of Lipschitz curves in a periodic cell Ω and $\varphi \in L^p(\Gamma, d\sigma)$ where $p > 1$ and σ is the line measure on Γ, we define

$$\mu(S) = \int_{S\cap\Gamma} \varphi\, d\sigma \tag{1.5}$$

for any Borel set $S \subset \Omega$. Extend μ to $\mathbb{R}^2$ by periodicity. Then μ satisfies (1.3) with $\varepsilon = 1 - (1/p) > 0$. It should be pointed out that the two dimensional case with a measure μ supported on a periodic system of piecewise smooth curves is studied in [**4**]. As the example above shows, our main theorem contains the result in [**4**] for $-\Delta + \mu$. We remark that the study in [**4**] of the Schrödinger operators with potentials supported on curves was motivated by a conjecture in [**11**] concerning the absolute continuity of the periodic Dirichlet-to-Neumann map in the photonic crystals theory. We refer the reader to [**11**] for an excellent survey on the mathematical analysis of photonic crystals.

In this paper we also study the spectrum of the elliptic operator $\mathcal{L} = \text{div}(\omega(\mathbf{x})\nabla)$ with a positive periodic scalar function $\omega(\mathbf{x})$. Note that, if ψ is an eigenfunction for $\mathcal{L}$, then $\varphi = \omega^{1/2}\psi$ is an eigenfunction for the operator $\omega(-\Delta + V)$ where

$$V = -\frac{1}{4}\left|\frac{\nabla\omega}{\omega}\right|^2 + \frac{\Delta\omega}{2\omega}. \tag{1.6}$$

It then follows from the uniform Sobolev inequalities in [**16**] that the spectrum of $\mathcal{L}$ is absolutely continuous if ω has second order partial derivatives in $L^{d/2}_{\text{loc}}$, $d \geq 3$. However, motivated by the methods used in [**19,22**], we are able to show that it suffices to assume $\omega \in C^1(\mathbb{R}^d)$.

Theorem 1.7. *Let $\omega(\mathbf{x})$ be a positive periodic function in $C^1(\mathbb{R}^d)$, $d \geq 2$ and $A = (a_{jl})_{d\times d}$ a symmetric, positive definite matrix with real constant entries. Then* *the spectrum of the operator* $\sum_{j,l} D_j(\omega(\mathbf{x})a_{jl}D_l)$ *on* $L^2(\mathbb{R}^d)$ *is purely absolutely continuous.*

The paper is organized as follows. In section 2 we give the definition of the self-adjoint operator $\mathbf{D}A\mathbf{D}^T + \mu$ on $L^2(\mathbb{R}^d)$. Our main theorem is proved in sections 3 and 4, while section 5 is devoted to the proof of Theorem 1.7. Finally, by a change of coordinates, we will assume that the measure μ and scalar function $\omega(\mathbf{x})$ are periodic with respect to the lattice $(2\pi\mathbb{Z})^d$ throughout the rest of the paper.

2. Some Preliminaries

We begin with a Poincaré inequality.

Lemma 2.1. *Let* Q *be a closed cube in* $\mathbb{R}^d$, $d \geq 2$ *with side length* $R \leq 1$. *Suppose that measure* μ *satisfies* $|\mu|(B(\mathbf{x}, r)) \leq C\, r^{d-2+\varepsilon}$ *for all* $\mathbf{x} \in Q$, $0 < r \leq 1$ *and some* $\varepsilon > 0$. *Then, for any* $\psi \in C^1(Q)$,

$$(2.2) \qquad \int_Q \int_Q |\psi(\mathbf{x}) - \psi(\mathbf{y})|^2\, d|\mu|(\mathbf{y}) d\mathbf{x} \leq C\, R^{d+\varepsilon} \int_Q |\nabla\psi(\mathbf{x})|^2 d\mathbf{x}.$$

Proof. See [**15, pp. 526-527**].

It follows from (2.2) that

$$(2.3) \qquad \int_Q |\psi|^2\, d|\mu| \leq C\, R^\varepsilon \int_Q |\nabla\psi|^2\, d\mathbf{x} + C\, R^{\varepsilon-2} \int_Q |\psi|^2\, d\mathbf{x}.$$

Lemma 2.4. *Suppose that* μ *satisfies (1.3) for some* $\varepsilon > 0$. *Then, for any* $\psi \in C^1(\mathbb{R}^d)$ *and* $\alpha > 0$,

$$(2.5) \qquad \int_{\mathbb{R}^d} |\psi|^2\, d|\mu| \leq \alpha \int_{\mathbb{R}^d} |\nabla\psi|^2\, d\mathbf{x} + C_\alpha \int_{\mathbb{R}^d} |\psi|^2\, d\mathbf{x}.$$

Proof. We divide $\mathbb{R}^d$ into a grid of cubes $\{Q_j\}$ with side length $R \leq 1$ and sum (2.3) over all Q_j. We obtain

$$(2.6) \qquad \int_{\mathbb{R}^d} |\psi|^2\, d|\mu| \leq C\, R^\varepsilon \int_{\mathbb{R}^d} |\nabla\psi|^2\, d\mathbf{x} + C\, R^{\varepsilon-2} \int_{\mathbb{R}^d} |\psi|^2\, d\mathbf{x}$$

for any $\psi \in C^1(\mathbb{R}^d)$. Estimate (2.5) then follows easily by choosing $R \in (0,1)$ such that $C\, R^\varepsilon \leq \alpha$.

It follows from (2.5) that the symmetric quadratic form on $L^2(\mathbb{R}^d)$ associated with $\sum_{j,l} D_j a_{jl} D_l + \mu$ with domain $H^1(\mathbb{R}^d)$ is semibounded and closed. Thus, by the KLMN Theorem [**14, vol.II**], it defines a unique self-adjoint operator which we denote by $\mathbf{D}A\mathbf{D}^T + \mu$.

Similarly, by dividing $\Omega = [0, 2\pi)^d \approx \mathbb{R}^d/(2\pi\mathbb{Z})^d \approx \mathbb{T}^d$ into a grid of cubes of side length $R \leq 1$, we obtain from (2.3) that

$$(2.7) \qquad \int_{\mathbb{T}^d} |\psi|^2\, d|\mu| \leq C\, R^\varepsilon \int_{\mathbb{T}^d} |\nabla\psi|^2\, d\mathbf{x} + C\, R^{\varepsilon-2} \int_{\mathbb{T}^d} |\psi|^2\, d\mathbf{x}.$$

This implies that, for any $\alpha > 0$,

$$(2.8) \qquad \int_{\mathbb{T}^d} |\psi|^2\, d|\mu| \leq \alpha \int_{\mathbb{T}^d} |\nabla\psi|^2\, d\mathbf{x} + C_\alpha \int_{\mathbb{T}^d} |\psi|^2\, d\mathbf{x}.$$

It follows that, for any $\mathbf{k} \in \mathbb{C}^d$, the quadratic form on $L^2(\mathbb{T}^d)$ associated with $\sum_{j,l}(D_j + k_j)a_{jl}(D_l + k_l) + \mu$ with domain $H^1(\mathbb{T}^d)$ is closed and thus define a unique closed operator, which we denote by $\mathcal{H}_\mu(\mathbf{k})$. Note that

$$\text{Domain}(\mathcal{H}_\mu(\mathbf{k})) = \{\psi \in H^1(\mathbb{T}^d) : \ \mathcal{H}_\mu(\mathbf{0})\psi \in L^2(\mathbb{T}^d)\} \tag{2.9}$$

is independent of $\mathbf{k} \in \mathbb{C}^d$. Moreover, for each $\mathbf{k} \in \mathbb{C}^d$, $\mathcal{H}_\mu(\mathbf{k})$ has compact resolvent.

As in the case of function potential $d\mu = V(\mathbf{x})d\mathbf{x}$ [**16**], we choose $\mathbf{a} = (a_1, \dots, a_d) \in \mathbb{R}^d$ such that

$$|\mathbf{a}| = 1, \quad \mathbf{a}A = (s_0, 0, \dots, 0), \quad \text{and} \quad s_0 > 0. \tag{2.10}$$

Let

$$L = \left\{\mathbf{b} \in \mathbb{R}^d : \ < \mathbf{b}, \mathbf{a} >= 0 \ \text{ and } |\mathbf{b}| \le \sqrt{d}\right\}. \tag{2.11}$$

Using the Floquet decomposition and the approach of Thomas [**21**], we have

Lemma 2.12. *If, for every $\mathbf{b} \in L$, the family of operators $\{\mathcal{H}_\mu((\delta + i\rho)\mathbf{a} + \mathbf{b}) : \ \rho \ge 1\}$ on $L^2(\mathbb{T}^d)$ with $\delta = \frac{1}{a_1}(\frac{1}{2} - b_1)$ has no common eigenvalue, then the spectrum of the operator $\mathbf{DAD}^T + \mu$ on $L^2(\mathbb{R}^d)$ is purely absolutely continuous.*

PROOF. See [**14, vol.IV**] and [**16**].

3. Estimate of $\mathcal{H}_\mu(\mathbf{k})$ on Sobolev Spaces with Negative Indices

The goal of this section is to prove the following theorem.

Theorem 3.1. *Let μ be a signed periodic Radon measure on $\mathbb{R}^d$, $d \ge 2$. Suppose that μ satisfies (1.3) for some $\varepsilon > 1$. Then the spectrum of $\mathbf{DAD}^T + \mu$ is purely absolutely continuous where $A = (a_{jl})_{d\times d}$ is a symmetric, positive definite matrix with real constant entries.*

To this end, we consider the Sobolev space on $\mathbb{T}^d$ of order $\gamma \in \mathbb{R}$,

$$H^\gamma(\mathbb{T}^d) = \left\{\psi : \ \psi(\mathbf{x}) = \sum_{\mathbf{n}\in\mathbb{Z}^d} a_{\mathbf{n}} e^{i\mathbf{n}\mathbf{x}} \quad \text{and} \quad \sum_{\mathbf{n}\in\mathbb{Z}^d} (1 + |\mathbf{n}|)^{2\gamma} |a_{\mathbf{n}}|^2 < \infty\right\} \tag{3.2}$$

with

$$\|\psi\|_{H^\gamma(\mathbb{T}^d)} = \left\{\sum_{\mathbf{n}\in\mathbb{Z}^d} (1 + |\mathbf{n}|)^{2\gamma} |a_{\mathbf{n}}|^2\right\}^{1/2}. \tag{3.3}$$

Lemma 3.4. *Suppose that μ satisfies (1.3) for some $\varepsilon \in (0, 2]$. Let $\gamma > 1 - (\varepsilon/2)$. Then, for any $\psi \in H^\gamma(\mathbb{T}^d)$,*

$$\left\{\int_{\mathbb{T}^d} |\psi|^2 \, d|\mu|\right\}^{1/2} \le C_\gamma \|\psi\|_{H^\gamma(\mathbb{T}^d)}. \tag{3.5}$$

PROOF. Let K_γ be the integrable function on $\mathbb{R}^d$ whose Fourier transform is $(1+|\xi|^2)^{-\gamma/2}$. We will show that

$$\left\{\int_{\mathbb{R}^d} |K_\gamma * f|^2 \, d|\mu|\right\}^{1/2} \leq C_\alpha \|f\|_{L^2(\mathbb{R}^d)}. \tag{3.6}$$

Assume (3.6) for a moment. Let $\psi \in H^\gamma(\mathbb{T}^d)$. We extend ψ to $\mathbb{R}^d$ by periodicity and consider $\widetilde{\psi} = \eta\psi$ where $\eta \in C_0^\infty(\mathbb{R}^d)$ and $\eta \equiv 1$ on $\Omega = [0, 2\pi)^d$. Using interpolation with nonnegative integers γ, we may show that

$$\|\widetilde{\psi}\|_{H^\gamma(\mathbb{R}^d)} \leq C_\gamma(\eta)\|\psi\|_{H^\gamma(\mathbb{T}^d)} \tag{3.7}$$

for any $\gamma > 0$. It follows that there exists $f \in L^2(\mathbb{R}^d)$ such that $\widetilde{\psi} = K_\gamma * f$ and $\|\widetilde{\psi}\|_{H^\gamma(\mathbb{R}^d)} \approx \|f\|_{L^2(\mathbb{R}^d)}$. Thus, by (3.6) and (3.7),

$$\begin{aligned}\left\{\int_\Omega |\psi|^2 \, d|\mu|\right\}^{1/2} &\leq \left\{\int_{\mathbb{R}^d} |\widetilde{\psi}|^2 d|\mu|\right\}^{1/2} = \left\{\int_{\mathbb{R}^d} |K_\gamma * f|^2 \, d|\mu|\right\}^{1/2} \\ &\leq C\|f\|_{L^2(\mathbb{R}^d)} \leq C\|\widetilde{\psi}\|_{H^\gamma(\mathbb{R}^d)} \leq C\|\psi\|_{H^\gamma(\mathbb{T}^d)}.\end{aligned}$$

To show (3.6), we use the following estimates on K_γ [**20, p.132**],

$$|K_\gamma(\mathbf{x})| \leq \begin{cases} \frac{C}{|\mathbf{x}|^{d-\gamma}}, & \text{if } |\mathbf{x}| \leq 1, \\ C\,e^{-c|\mathbf{x}|}, & \text{if } |\mathbf{x}| < 1. \end{cases} \tag{3.8}$$

Thus,

$$\begin{aligned}|(K_\gamma * f)(\mathbf{x})| &\leq C\int_{|\mathbf{y}-\mathbf{x}|\leq 1} \frac{|f(\mathbf{y})|\, d\mathbf{y}}{|\mathbf{y}-\mathbf{x}|^{d-\gamma}} + C\int_{|\mathbf{y}-\mathbf{x}|>1} e^{-c|\mathbf{y}-\mathbf{x}|}|f(\mathbf{y})|\, d\mathbf{y} \\ &= C\left\{I(f)(\mathbf{x}) + J(f)(\mathbf{x})\right\}.\end{aligned}$$

Since

$$J(f)(\mathbf{x}) \leq C\left\{\int_{\mathbb{R}^d} e^{-c|\mathbf{y}-\mathbf{x}|}|f(\mathbf{y})|^2\, d\mathbf{y}\right\}^{1/2}$$

and $|\mu|$ is periodic, it is easy to see that

$$\left\{\int_{\mathbb{R}^d} |J(f)|^2 \, d|\mu|\right\}^{1/2} \leq C\|f\|_{L^2(\mathbb{R}^d)}. \tag{3.9}$$

To estimate $I(f)(\mathbf{x})$, we write

$$I(f)(\mathbf{x}) = \sum_{j=0}^\infty \int_{2^{-j-1}<|\mathbf{y}-\mathbf{x}|\leq 2^{-j}} \frac{|f(\mathbf{y})|\, d\mathbf{y}}{|\mathbf{y}-\mathbf{x}|^{d-\gamma}} = \sum_{j=0}^\infty T_j(f)(\mathbf{x}).$$

Clearly,

$$\|T_j(f)\|_{L^\infty(\mathbb{R}^d, d|\mu|)} \leq C\,(2^{-j})^\gamma \|f\|_{L^\infty(\mathbb{R}^d, d\mathbf{x})}. \tag{3.10}$$

Also,

$$\begin{aligned}\|T_j(f)\|_{L^1(\mathbb{R}^d,d|\mu|)} &\le C(2^j)^{d-\gamma}\int_{\mathbb{R}^d}\left\{\int_{|\mathbf{y}-\mathbf{x}|\le 2^{-j}}|f(\mathbf{y})|\,d\mathbf{y}\right\}d|\mu|(\mathbf{x})\\ &= C\,(2^j)^{d-\gamma}\int_{\mathbb{R}^d}|f(\mathbf{y})|\,|\mu|(B(\mathbf{y},2^{-j}))\,d\mathbf{y}\\ &\le C(2^{-j})^{\gamma+\varepsilon-2}\|f\|_{L^1(\mathbb{R}^d,d\mathbf{x})}.\end{aligned}$$

This, together with (3.10), implies that

$$\|T_j(f)\|_{L^2(\mathbb{R}^d,d|\mu|)} \le C\,(2^{-j})^{\gamma+(\varepsilon/2)-1}\|f\|_{L^2(\mathbb{R}^d,d\mathbf{x})} \tag{3.11}$$

by interpolation. Thus, if $\gamma > 1-(\varepsilon/2)$,

$$\|I(f)\|_{L^2(\mathbb{R}^d,d|\mu|)} \le \sum_{j=0}^{\infty}\|T_j(f)\|_{L^2(\mathbb{R}^d,d|\mu|)} \le C\|f\|_{L^2(\mathbb{R}^d,d\mathbf{x})}. \tag{3.12}$$

Estimate (3.6) now follows from (3.9) and (3.12).

Lemma 3.13. *Suppose that μ satisfies (1.3) for some $\varepsilon \in (0,2]$. Then, for any $\gamma > 1-(\varepsilon/2)$ and $\psi \in H^\gamma(\mathbb{T}^d)$,*

$$\|\psi\mu\|_{H^{-\gamma}(\mathbb{T}^d)} \le C_\gamma\|\psi\|_{H^\gamma(\mathbb{T}^d)}. \tag{3.14}$$

Proof. Let $\phi \in H^\gamma(\mathbb{T}^d)$. By the Cauchy inequality and (3.5),

$$\begin{aligned}\left|\int_{\mathbb{T}^d}\psi\phi\,d\mu\right| &\le \int_{\mathbb{T}^d}|\psi|\,|\phi|\,d|\mu|\\ &\le \left\{\int_{\mathbb{T}^d}|\psi|^2\,d|\mu|\right\}^{1/2}\left\{\int_{\mathbb{T}^d}|\phi|^2\,d|\mu|\right\}^{1/2}\\ &\le C_\gamma\|\psi\|_{H^\gamma(\mathbb{T}^d)}\|\phi\|_{H^\gamma(\mathbb{T}^d)}.\end{aligned}$$

Estimate (3.14) then follows since $H^{-\gamma}(\mathbb{T}^d)$ is the dual space of $H^\gamma(\mathbb{T}^d)$.

Lemma 3.15. *Let $\mathbf{k} = (\delta + i\rho)\mathbf{a} + \mathbf{b}$ where $\mathbf{a}$ satisfies (2.10), $\mathbf{b} \in L$, $\rho \ge 1$, and $\delta = \frac{1}{a_1}(\frac{1}{2}-b_1)$. Then, for $\gamma \in [0,1/2]$,*

$$\rho^{1-2\gamma}\|\psi\|_{H^\gamma(\mathbb{T}^d)} \le C_\gamma\|(\mathbf{D}+\mathbf{k})A(\mathbf{D}+\mathbf{k})^T\psi\|_{H^{-\gamma}(\mathbb{T}^d)} \tag{3.16}$$

where C_γ does not depend on ρ.

Proof. It suffices to show that, for any $\mathbf{n} \in \mathbb{Z}^d$,

$$|(\mathbf{n}+\mathbf{k})A(\mathbf{n}+\mathbf{k})^T| \ge c\,(|\mathbf{n}|+\rho). \tag{3.17}$$

This would imply that

$$|(\mathbf{n}+\mathbf{k})A(\mathbf{n}+\mathbf{k})^T| \ge C_\gamma|\mathbf{n}|^{2\gamma}\rho^{1-2\gamma} \quad \text{for any } 0 \le \gamma \le 1/2. \tag{3.18}$$

Estimate (3.16) follows easily from (3.18).

To see (3.17), first note that

$$(\mathbf{n}+\mathbf{k})A(\mathbf{n}+\mathbf{k})^T = (\mathbf{n}+\delta\mathbf{a}+\mathbf{b})A(\mathbf{n}+\delta\mathbf{a}+\mathbf{b})^T - \rho^2 s_0 a_1 + 2i\rho s_0(n_1+\delta a_1+b_1).$$

Thus

$$|(\mathbf{n}+\mathbf{k})A(\mathbf{n}+\mathbf{k})^T| \geq 2\rho s_0|n_1+\delta a_1+b_1| \geq 2\rho s_0|n_1+\frac{1}{2}| \geq \rho s_0. \tag{3.19}$$

Also,

$$c|\mathbf{n}+\mathbf{k}|^2 \leq |(\mathbf{n}+\mathbf{k})A(\mathbf{n}+\bar{\mathbf{k}})^T| \leq |(\mathbf{n}+\mathbf{k})A(\mathbf{n}+\mathbf{k})^T| + C\rho|\mathbf{n}+\mathbf{k}|$$

where $\bar{\mathbf{k}}$ denotes the complex conjugate of $\mathbf{k}\in\mathbb{C}^d$. It follows that

$$|\mathbf{n}+\mathbf{k}|^2 \leq C\,|(\mathbf{n}+\mathbf{k})A(\mathbf{n}+\mathbf{k})^T| + C\rho^2 \leq C|(\mathbf{n}+\mathbf{k})A(\mathbf{n}+\mathbf{k})^T|^2.$$

Hence,

$$|\mathbf{n}| \leq |\mathbf{n}+\mathbf{k}| + C\rho \leq C|(\mathbf{n}+\mathbf{k})A(\mathbf{n}+\mathbf{k})^T|.$$

This, together with (3.19), yields (3.17).

We are now in a position to give the

Proof of Theorem 3.1.

By Lemma 2.12, it suffices to show that the family of operators $\{\mathcal{H}_\mu((\delta+i\rho)\mathbf{a}+\mathbf{b}):\ \rho\geq 1\}$ on $L^2(\mathbb{T}^d)$ has no common eigenvalue.

Let $\mathbf{k}=(\delta+i\rho)\mathbf{a}+\mathbf{b}\in\mathbb{C}^d$. Fix $E\in\mathbb{R}$. Suppose, for each $\rho\geq 1$, there exists $\psi_\rho\in$Domain$(\mathcal{H}_\mu(\mathbf{k}))$ such that $\|\psi_\rho\|_{L^2(\mathbb{T}^d)}=1$ and

$$\left\{(\mathbf{D}+\mathbf{k})A(\mathbf{D}+\mathbf{k})^T+\mu\right\}\psi_\rho = E\psi_\rho.$$

Since $\psi_\rho\in H^1(\mathbb{T}^d)$, by Lemma 3.13, $\mu\psi_\rho\in H^{-\gamma}(\mathbb{T}^d)$ for any $\gamma\in(1-(\varepsilon/2),1]$. Since $\varepsilon\in(1,2]$, we may choose γ so that $1-(\varepsilon/2)<\gamma<1/2$. It follows from Lemmas 3.15 and 3.13 that

$$\begin{aligned}\rho^{1-2\gamma}\|\psi_\rho\|_{H^\gamma(\mathbb{T}^d)} &\leq C\|(\mathbf{D}+\mathbf{k})A(\mathbf{D}+\mathbf{k})^T\psi_\rho\|_{H^{-\gamma}(\mathbb{T}^d)}\\ &\leq C\,|E|\,\|\psi_\rho\|_{H^{-\gamma}(\mathbb{T}^d)} + C\,\|\mu\psi_\rho\|_{H^{-\gamma}(\mathbb{T}^d)}\\ &\leq C\,|E|\,\|\psi_\rho\|_{H^{-\gamma}(\mathbb{T}^d)} + C\,\|\psi_\rho\|_{H^{\gamma}(\mathbb{T}^d)}\\ &\leq \{C\,|E|+C\}\|\psi_\rho\|_{H^\gamma(\mathbb{T}^d)}.\end{aligned}$$

This implies that

$$\rho^{1-2\gamma} \leq C\,|E| + C$$

for some $\gamma<1/2$, which is impossible if we let $\rho\to\infty$.

4. A Stronger Result for the Two-Dimensional Case

In this section we will show that it suffices to assume $\varepsilon>0$ in Theorem 3.1 in the case $d=2$.

Theorem 4.1. *Let μ be a signed periodic Radon measure on $\mathbb{R}^2$. Suppose that μ satisfies (1.3) for some $\varepsilon>0$. Then the spectrum of $\mathbf{D}A\mathbf{D}^T+\mu$ is purely absolutely continuous where $A=(a_{jl})_{2\times 2}$ is a symmetric, positive definite matrix with real constant entries.*

To prove Theorem 4.1, we consider the space $\mathcal{M}(\mathbb{T}^d)$ of complex measures on $\mathbb{T}^d$. More precisely, $\mathcal{M}(\mathbb{T}^d)$ consists of all periodic complex measures on $\mathbb{R}^d$ with periodic cell $\Omega=[0,2\pi)^d$ and

$$\|\nu\|_{\mathcal{M}(\mathbb{T}^d)} = |\nu|(\Omega). \tag{4.2}$$

For $\nu \in \mathcal{M}(\mathbb{T}^d)$, we define

$$\hat{\nu}(\mathbf{n}) = \frac{1}{(2\pi)^d} \int_\Omega e^{-i\mathbf{n}\mathbf{y}} \, d\nu(\mathbf{y}). \tag{4.3}$$

Let $\mathbf{k} = (\delta + i\rho)\mathbf{a} + \mathbf{b}$ as in Lemma 3.15 and

$$G_\rho(\mathbf{x}) = \frac{1}{(2\pi)^d} \sum_{\mathbf{n}\in\mathbb{Z}^d} \frac{e^{i\mathbf{n}\mathbf{x}}}{(\mathbf{n}+\mathbf{k})A(\mathbf{n}+\mathbf{k})^T}. \tag{4.4}$$

Note that $G_\rho \in L^2(\mathbb{T}^d)$ if $d = 2$ or 3. The following pointwise estimate was established in [**18**],

$$|G_\rho(\mathbf{x})| \le \begin{cases} \frac{C}{\operatorname{dist}(\mathbf{x},(2\pi\mathbb{Z})^d)}, & \text{if } d = 3, \\ C \ln\left[1 + \frac{1}{\operatorname{dist}(\mathbf{x},(2\pi\mathbb{Z})^d)}\right], & \text{if } d = 2. \end{cases} \tag{4.5}$$

Lemma 4.6. *Let $\nu \in \mathcal{M}(\mathbb{T}^d)$ with $d = 2$ or 3. Define*

$$\psi(\mathbf{x}) = \sum_{\mathbf{n}\in\mathbb{Z}^d} \frac{\hat{\nu}(\mathbf{n}) e^{i\mathbf{n}\mathbf{x}}}{(\mathbf{n}+\mathbf{k})A(\mathbf{n}+\mathbf{k})^T}. \tag{4.7}$$

Suppose $\psi \in H^1(\mathbb{T}^d)$. Then

$$\psi(\mathbf{x}) = \int_\Omega G_\rho(\mathbf{x}-\mathbf{y}) \, d\nu(\mathbf{y}) \tag{4.8}$$

for a.e. $\mathbf{x} \in \Omega$ with respect to the measure $|\mu|$.

Proof. We give the proof for the case $d = 3$. The two dimensional case may be handled in the same manner.

Let $g \in \mathcal{S}(\mathbb{R}^d)$ such that its Fourier transform $\hat{g}(\xi) = 1$ for $|\xi| \le 1/2$, and $\hat{g}(\xi) = 0$ for $|\xi| \ge 1$. We first claim that it suffices to prove that (4.8) holds for each $\mathbf{x} \in \mathbb{T}^d$ with the properties

$$\int_\Omega \frac{d|\nu|(\mathbf{y})}{|\mathbf{x}-\mathbf{y}|} < \infty \tag{4.9}$$

and

$$\sum_{\mathbf{n}\in\mathbb{Z}^d} \frac{\hat{g}(\frac{\mathbf{n}}{N})\hat{\nu}(\mathbf{n}) e^{i\mathbf{n}\mathbf{x}}}{(\mathbf{n}+\mathbf{k})A(\mathbf{n}+\mathbf{k})^T} \quad \text{converges to } \psi(\mathbf{x}) \tag{4.10}$$

for a sequence $N = N_j \to \infty$. Indeed, by condition (1.3) for $\varepsilon > 0$,

$$\sup_{\mathbf{y}\in\Omega} \int_\Omega \frac{d|\mu|(\mathbf{x})}{|\mathbf{x}-\mathbf{y}|} < \infty.$$

It follows that

$$\int_\Omega \left\{ \int_\Omega \frac{d|\nu|(\mathbf{y})}{|\mathbf{x}-\mathbf{y}|} \right\} d|\mu|(\mathbf{x}) \le C \int_\Omega d|\nu|(\mathbf{y}) < \infty.$$

This implies that (4.9) holds for a.e. $\mathbf{x} \in \Omega$ with respect to $|\mu|$. Also, since $\psi \in H^1(\mathbb{T}^d)$, the sum in (4.10) converges to ψ in $H^1(\mathbb{T}^d)$ as $N \to \infty$. By (2.8), this implies that the sum in (4.10) converges to ψ in $L^2(\Omega, d|\mu|)$. It follows that there exists a sequence $\{N_j\}$ which goes to infinite such that the sum converges pointwise to $\psi(\mathbf{x})$ as $N = N_j \to \infty$ for a.e. $\mathbf{x} \in \Omega$ with respect to $|\mu|$. Thus (4.9) and (4.10) hold for a.e. $\mathbf{x} \in \Omega$ with respect to $|\mu|$.

We now fix $\mathbf{x}_0 \in \Omega$ with properties (4.9) and (4.10). Note that

$$\begin{aligned}\psi(\mathbf{x}_0) &= \lim_{j\to\infty} \sum_{\mathbf{n}\in\mathbb{Z}^d} \frac{\hat{g}(\frac{\mathbf{n}}{N_j})\hat{\nu}(\mathbf{n})e^{i\mathbf{n}\mathbf{x}_0}}{(\mathbf{n}+\mathbf{k})A(\mathbf{n}+\mathbf{k})^T} \\ &= \lim_{j\to\infty} \frac{1}{(2\pi)^d} \int_\Omega \sum_{\mathbf{n}\in\mathbb{Z}^d} \frac{\hat{g}(\frac{\mathbf{n}}{N_j})e^{i\mathbf{n}(\mathbf{x}_0-\mathbf{y})}}{(\mathbf{n}+\mathbf{k})A(\mathbf{n}+\mathbf{k})^T} d\nu(\mathbf{y}) \\ &= \lim_{j\to\infty} \int_\Omega G_{\rho,N_j}(\mathbf{x}_0-\mathbf{y})d\nu(\mathbf{y}),\end{aligned}$$

where

$$G_{\rho,N_j}(\mathbf{x}) = \frac{1}{(2\pi)^d} \sum_{\mathbf{n}\in\mathbb{Z}^d} \frac{\hat{g}(\frac{\mathbf{n}}{N_j})e^{i\mathbf{n}\mathbf{x}}}{(\mathbf{n}+\mathbf{k})A(\mathbf{n}+\mathbf{k})^T} = \int_\Omega h_j(\mathbf{x}-\mathbf{y})G_\rho(\mathbf{y})d\mathbf{y} \tag{4.11}$$

and

$$h_j(\mathbf{x}) = \sum_{\mathbf{n}\in\mathbb{Z}^d} \hat{g}(\frac{\mathbf{n}}{N_j})e^{i\mathbf{n}\mathbf{x}} = \frac{1}{(2\pi)^d} \sum_{\mathbf{n}\in\mathbb{Z}^d} (N_j)^d g(N_j(\mathbf{x}+2\pi\mathbf{n})) \tag{4.12}$$

by Poisson summation formula.

Since $g \in \mathcal{S}(\mathbb{R}^d)$, it is easy to see that

$$|h_j(\mathbf{x})| \le \frac{C N_j^d}{\{1+N_j \mathrm{dist}(\mathbf{x},(2\pi\mathbb{Z})^d)\}^{d+1}} \qquad \text{for } \mathbf{x}\in\Omega. \tag{4.13}$$

It follows that, for $\mathbf{x} \in \Omega$,

$$\begin{aligned}|G_{\rho,N_j}(\mathbf{x})| &\le \int_\Omega |h_j(\mathbf{x}-\mathbf{y})|\,|G_\rho(\mathbf{y})|\,d\mathbf{y} \\ &\le C\int_\Omega \frac{N_j^d}{\{1+N_j\mathrm{dist}(\mathbf{x}-\mathbf{y},(2\pi\mathbb{Z})^d)\}^{d+1}} \cdot \frac{1}{\mathrm{dist}(\mathbf{y},(2\pi\mathbb{Z})^d)} d\mathbf{y} \\ &\le \frac{C}{\mathrm{dist}(\mathbf{x},(2\pi\mathbb{Z})^d)}.\end{aligned}$$

Using the fact that $G_\rho \in C^\infty(\mathbb{R}^d \setminus (2\pi\mathbb{Z})^d)$ and estimate (4.13), one may show that $G_{\rho,N_j}(\mathbf{x})$ converges to $G_\rho(\mathbf{x})$ uniformly on $\{\mathbf{x}\in\Omega : \mathrm{dist}(\mathbf{x},(2\pi\mathbb{Z})^d) \ge \alpha\}$ for any $\alpha > 0$. It follows that

$$\begin{aligned}&\limsup_{j\to\infty} \Big| \int_\Omega G_{\rho,N_j}(\mathbf{x}_0-\mathbf{y})d\nu(\mathbf{y}) - \int_\Omega G_\rho(\mathbf{x}_0-\mathbf{y})\,d\nu(\mathbf{y})\Big| \\ &\le C\int_{\{\mathbf{y}\in\Omega:\ \mathrm{dist}(\mathbf{x}_0-\mathbf{y},(2\pi\mathbb{Z})^d)<\alpha\}} \frac{d|\nu|(\mathbf{y})}{\mathrm{dist}(\mathbf{x}_0-\mathbf{y},(2\pi\mathbb{Z})^d)}\end{aligned}$$

for any $\alpha > 0$. By (4.9) this implies that

$$\psi(\mathbf{x}_0) = \lim_{j\to\infty} \int_\Omega G_{\rho,N_j}(\mathbf{x}_0-\mathbf{y})\,d\nu(\mathbf{y}) = \int_\Omega G_\rho(\mathbf{x}_0-\mathbf{y})\,d\nu(\mathbf{y}).$$

The proof is complete.

Remark 4.14. Since the Lebesgue measure on $\mathbb{R}^d$ satisfies (1.3) with $\varepsilon = 2$, it follows from the proof of Lemma 4.6 that (4.8) also holds for a.e. $\mathbf{x} \in \Omega$ with respect to the Lebesgue measure.

We need one more lemma before we carry out the proof of Theorem 4.1.

Lemma 4.15. *Let μ be a signed periodic Radon measure on $\mathbb{R}^2$. Suppose that μ satisfies (1.3) for some $\varepsilon \in (0,1)$. Then*

$$\lim_{\rho\to\infty} \sup_{\mathbf{x}\in\Omega} \int_\Omega |G_\rho(\mathbf{x}-\mathbf{y})|\, d|\mu|(\mathbf{y}) = 0.$$

PROOF. We will show that, for any $\alpha \in (0,1)$, we have

$$|G_\rho(\mathbf{x})| \le \alpha + \frac{C_\alpha}{\rho} + C_\alpha \ln\left\{1 + \frac{1}{\rho\,\mathrm{dist}(\mathbf{x},(2\pi\mathbb{Z})^2)}\right\}. \tag{4.16}$$

Since

$$\begin{aligned}\int_\Omega \ln\left\{1+\frac{1}{\rho|\mathbf{x}|}\right\} d|\mu|(\mathbf{x}) &= \int_0^\infty |\mu|\left\{\mathbf{x}\in\Omega:\ \ln\left\{1+\frac{1}{\rho|\mathbf{x}|}\right\} > \lambda\right\} d\lambda \\ &\le \frac{C}{\rho^\varepsilon}\int_0^\infty \frac{d\lambda}{(e^\lambda-1)^\varepsilon} \\ &\le \frac{C_\varepsilon}{\rho^\varepsilon}\end{aligned}$$

by (1.3), Lemma 4.15 follows easily from estimate (4.16).

Estimate (4.16) follows by inspecting the proof of (4.6) in [**18**]. Indeed, as in [**18**], we write

$$G_\rho(\mathbf{x}) = \frac{1}{(2\pi)^2}\sum_{\mathbf{n}\in\mathbb{Z}^2} e^{-i(\delta\mathbf{a}+\mathbf{b})(\mathbf{x}+2\pi\mathbf{n})} F_\rho(\mathbf{x}+2\pi\mathbf{n}) \tag{4.17}$$

where F_ρ is the inverse Fourier transform of

$$\frac{\eta\left(\frac{|\varphi(\xi,\rho)|}{\rho^2}\right)}{\varphi(\xi,\rho)}, \tag{4.18}$$

$\varphi(\xi,\rho) = \xi A\xi^T - \rho^2 s_0 a_1 + 2i\rho s_0 \xi_1$ and η is a function in $C^\infty(\mathbb{R}_+)$ such that $\eta(r) = 1$ if $r \ge s_0^2$ and $\eta(r) = 0$ if $0 < r < s_0^2/2$. Thus

$$|G_\rho(\mathbf{x})| \le C \sum_{\mathbf{n}\in\mathbb{Z}^2} |F_\rho(\mathbf{x}+2\pi\mathbf{n})|. \tag{4.19}$$

By estimates (2.24) and (3.25) in [**18**], we have

$$\begin{aligned}|F_\rho(\mathbf{x})| &\le \frac{C}{|\mathbf{x}|^4},\\ |F_\rho(\mathbf{x})| &\le C\left\{\frac{1}{\rho} + \ln\left(1+\frac{1}{\rho|\mathbf{x}|}\right)\right\}.\end{aligned} \tag{4.20}$$

This implies that

$$|G_\rho(\mathbf{x})| \le C \sum_{\substack{\mathbf{n}\in\mathbb{Z}^2\\ |\mathbf{x}+2\pi\mathbf{n}|\ge N}} \frac{1}{|\mathbf{x}+2\pi\mathbf{n}|^4} + C \sum_{\substack{\mathbf{n}\in\mathbb{Z}^2\\ |\mathbf{x}+2\pi\mathbf{n}|< N}} \left\{\frac{1}{\rho} + \ln\left(1+\frac{1}{\rho|\mathbf{x}+2\pi\mathbf{n}|}\right)\right\}.$$

From this, estimate (4.16) follows by choosing N large.

We are now ready to give the

Proof of Theorem 4.1.

Let $\mathbf{k} = (\delta + i\rho)\mathbf{a} + \mathbf{b} \in \mathbb{C}^2$. By Lemma 2.12, it suffices to show that the family of operators $\{\mathcal{H}_\mu(\mathbf{k}) : \rho \geq 1\}$ on $L^2(\mathbb{T}^2)$ has no common eigenvalue.

To this end, we fix $E \in \mathbb{R}$. Suppose that, for each $\rho \geq 1$, there exists $\psi_\rho \in$Domain$(\mathcal{H}_\mu(\mathbf{k}))$ such that $\|\psi_\rho\|_{L^2(\mathbb{T}^2)} = 1$ and

$$\{(\mathbf{D}+\mathbf{k})A(\mathbf{D}+\mathbf{k})^T + \mu\}\,\psi_\rho = E\psi_\rho. \tag{4.21}$$

Since $\mu\psi_\rho \in \mathcal{M}(\mathbb{T}^2)$, we have $(\mathbf{D}+\mathbf{k})A(\mathbf{D}+\mathbf{k})^T\psi_\rho \in \mathcal{M}(\mathbb{T}^2)$. Let

$$\nu_\rho = (\mathbf{D}+\mathbf{k})A(\mathbf{D}+\mathbf{k})^T\psi_\rho.$$

Then

$$\psi_\rho(\mathbf{x}) = \sum_{\mathbf{n}\in\mathbb{Z}^2} \frac{\hat{\nu_\rho}\, e^{i\mathbf{n}\mathbf{x}}}{(\mathbf{n}+\mathbf{k})A(\mathbf{n}+\mathbf{k})^T}. \tag{4.22}$$

Since $\psi_\rho \in H^1(\mathbb{T}^2)$, in view of Lemma 4.6, we have

$$|\psi_\rho(\mathbf{x})| \leq \int_\Omega |G_\rho(\mathbf{x}-\mathbf{y})|\, d|\nu_\rho|(\mathbf{y}) \tag{4.23}$$

for a.e. $\mathbf{x} \in \Omega$ with respect to $|\mu|$. Hence, by (4.21),

$$\begin{aligned}
\|\nu_\rho\|_{\mathcal{M}(\mathbb{T}^2)} &\leq |E|\,\|\psi_\rho\|_{L^1(\mathbb{T}^2,d\mathbf{x})} + \int_\Omega |\psi_\rho|\, d|\mu| \\
&\leq |E|\,\|\psi_\rho\|_{L^1(\mathbb{T}^2,d\mathbf{x})} + \int_\Omega \left\{\int_\Omega |G_\rho(\mathbf{x}-\mathbf{y})|\, d|\nu_\rho|(\mathbf{y})\right\} d|\mu|(\mathbf{x}) \\
&= |E|\,\|\psi_\rho\|_{L^1(\mathbb{T}^2,d\mathbf{x})} + \int_\Omega \left\{\int_\Omega |G_\rho(\mathbf{x}-\mathbf{y})|\, d|\mu|(\mathbf{x})\right\} d|\nu_\rho|(\mathbf{y}).
\end{aligned}$$

By Lemma 4.15, this implies that

$$\|\nu_\rho\|_{\mathcal{M}(\mathbb{T}^2)} \leq |E|\,\|\psi_\rho\|_{L^1(\mathbb{T}^2,d\mathbf{x})} + \frac{1}{2}\|\nu_\rho\|_{\mathcal{M}(\mathbb{T}^2)}$$

for ρ sufficiently large. Consequently,

$$\|\nu_\rho\|_{\mathcal{M}(\mathbb{T}^2)} \leq 2|E|\|\psi_\rho\|_{L^1(\mathbb{T}^2,d\mathbf{x})} \qquad \text{for } \rho \text{ large.} \tag{4.24}$$

We claim that (4.24) is impossible. Indeed, by Remark 4.14, (4.23) holds for a.e. $\mathbf{x} \in \Omega$ with respect to the Lebesgue measure on $\mathbb{R}^2$. It follows that

$$\begin{aligned}
\|\psi_\rho\|_{L^1(\mathbb{T}^2,d\mathbf{x})} &\leq \int_\Omega \left\{\int_\Omega |G_\rho(\mathbf{x}-\mathbf{y})|\, d\mathbf{x}\right\} d|\nu_\rho|(\mathbf{y}) \\
&\leq \frac{C}{\rho^{1/2}}\,\|\nu_\rho\|_{\mathcal{M}(\mathbb{T}^2)}
\end{aligned}$$

where we have used the estimate

$$\int_\Omega |G_\rho(\mathbf{x})|\, d\mathbf{x} \leq \frac{C}{\rho^{1/2}}$$

(see (4.7) in [**18**]). This contradicts with (4.24). The proof is finished.

5. The Absolute Continuity of $\mathbf{div}(\omega(\mathbf{x})\nabla)$

In this section we give the proof of Theorem 1.7.

Let ω be a positive periodic function on $\mathbb{R}^d$, $d \geq 2$ and $A = (a_{jl})_{d\times d}$ a symmetric, positive definite matrix with real constant entries. Suppose $\omega \in C^1(\mathbb{R}^d)$. We consider the self-adjoint operator on $L^2(\mathbb{R}^d)$,

$$\mathcal{L} = \sum_{j,l} D_j(\omega(\mathbf{x})a_{jl}D_l), \tag{5.1}$$

defined by its associated quadratic form. Without the loss of generality, we assume that ω is periodic with respect to the lattice $(2\pi\mathbb{Z})^d$.

As in section 2, let $\mathbf{k} = (\delta + i\rho)\mathbf{a} + \mathbf{b} \in \mathbb{C}^d$. Consider the closed operator on $L^2(\mathbb{T}^d)$,

$$\mathcal{L}_\omega(\mathbf{k}) = \sum_{j,l}(D_j + k_j)(\omega(\mathbf{x})a_{jl}(D_l + k_l)), \tag{5.2}$$

defined by its associated quadratic form. Note that, since $\omega \in C^1(\mathbb{T}^d)$,

$$\text{Domain}(\mathcal{L}_\omega(\mathbf{k})) = \left\{ \psi \in H^1(\mathbb{T}^d) : \sum_{j,l} D_j(\omega(\mathbf{x})a_{jl}D_l\psi) \in L^2(\mathbb{T}^d) \right\}.$$

We will show that the family of operators $\{\mathcal{L}_\omega(\mathbf{k}) : \rho \geq 1\}$ on $L^2(\mathbb{T}^d)$ has no common eigenvalue. By the Thomas approach, this implies Theorem 1.7.

To this end, we fix $E \in \mathbb{R}$. Suppose that, for each $\rho \geq 1$, there exists $\psi_\rho \in$Domain$(\mathcal{L}_\omega(\mathbf{k}))$ such that

$$\|\psi_\rho\|_{L^2(\mathbb{T}^d)} = 1 \quad \text{and} \quad \mathcal{L}_\omega(\mathbf{k})\psi_\rho = E\psi_\rho. \tag{5.3}$$

Let

$$\varphi_\rho = \widetilde{\omega}^{1/2}\psi_\rho \tag{5.4}$$

where $\widetilde{\omega}$ is a positive function in $C^2(\mathbb{T}^d)$. Clearly, $\varphi_\rho \in H^1(\mathbb{T}^d)$. A computation shows that

$$\begin{aligned}
&(\mathbf{D}+\mathbf{k})A(\mathbf{D}+\mathbf{k})^T\varphi_\rho + \sum_{j,l} a_{jl}D_j\big(\ln(\frac{\omega}{\widetilde{\omega}})\big)(D_l + k_l)\varphi_\rho\\
&-\left\{\frac{1}{2\widetilde{\omega}}\sum_{j,l} a_{jl}D_jD_l\widetilde{\omega} + \frac{1}{2\omega\widetilde{\omega}}\sum_{j,l} a_{jl}(D_j\omega)(D_l\widetilde{\omega}) - \frac{3}{4\widetilde{\omega}^2}\sum_{j,l} a_{jl}(D_j\widetilde{\omega})(D_l\widetilde{\omega})\right\}\varphi_\rho\\
&\qquad = \frac{E\varphi_\rho}{\omega}.
\end{aligned}$$

This implies that $(\mathbf{D}+\mathbf{k})A(\mathbf{D}+\mathbf{k})^T\varphi_\rho \in L^2(\mathbb{T}^d)$ and

$$\begin{aligned}
&\|(\mathbf{D}+\mathbf{k})A(\mathbf{D}+\mathbf{k})^T\varphi_\rho\|_{L^2(\mathbb{T}^d)} \leq C\,\|\mathbf{D}\big(\ln(\frac{\omega}{\widetilde{\omega}})\big)\|_{C(\mathbb{T}^d)}\|(\mathbf{D}+\mathbf{k})\varphi_\rho\|_{L^2(\mathbb{T}^d)}\\
&\quad + C\left\{\|\frac{\mathbf{D}A\mathbf{D}^T\widetilde{\omega}}{\widetilde{\omega}}\|_{C(\mathbb{T}^d)} + \|\frac{\mathbf{D}\omega}{\omega}\|_{C(\mathbb{T}^d)}\|\frac{\mathbf{D}\widetilde{\omega}}{\widetilde{\omega}}\|_{C(\mathbb{T}^d)} + \|\frac{\mathbf{D}\widetilde{\omega}}{\widetilde{\omega}}\|^2_{C(\mathbb{T}^d)}\right\}\|\varphi_\rho\|_{L^2(\mathbb{T}^d)}\\
&\quad + |E|\,\|\frac{1}{\omega}\|_{C(\mathbb{T}^d)}\|\varphi_\rho\|_{L^2(\mathbb{T}^d)}.
\end{aligned} \tag{5.5}$$

By (3.17),

$$\rho\|\varphi_\rho\|_{L^2(\mathbb{T}^d)} + \|(\mathbf{D}+\mathbf{k})\varphi_\rho\|_{L^2(\mathbb{T}^d)} \le C\|(\mathbf{D}+\mathbf{k})A(\mathbf{D}+\mathbf{k})^T\varphi_\rho\|_{L^2(\mathbb{T}^d)}. \tag{5.6}$$

It follows from (5.5) and (5.6) that

$$\begin{aligned}\Big\{1 - C\|\mathbf{D}\big(\ln(\frac{\omega}{\widetilde{\omega}})\big)\|_{C(\mathbb{T}^d)}\Big\}\, &\|(\mathbf{D}+\mathbf{k})A(\mathbf{D}+\mathbf{k})^T\varphi_\rho\|_{L^2(\mathbb{T}^d)} \\ &\le C(\omega,\widetilde{\omega},|E|)\,\|\varphi_\rho\|_{L^2(\mathbb{T}^d)} \\ &\le \widetilde{C}(\omega,\widetilde{\omega},|E|)\end{aligned}$$

where $\widetilde{C}(\omega,\widetilde{\omega},|E|)$ does not depend on $\rho \ge 1$. In view of (5.6), this is impossible if we can find a positive function $\widetilde{\omega} \in C^2(\mathbb{T}^d)$ such that

$$C\,\|\mathbf{D}\big(\ln(\frac{\omega}{\widetilde{\omega}})\big)\|_{C(\mathbb{T}^d)} \le \frac{1}{2}. \tag{5.7}$$

To see (5.7), let $g \in C_0^\infty(B(\mathbf{0},1))$ such that $g \ge 0$ and $\int_{\mathbb{R}^d} g(\mathbf{x})\,d\mathbf{x} = 1$. For $0 < \alpha < 1$, define

$$h_\alpha(\mathbf{x}) = \frac{1}{\alpha^d}\sum_{\mathbf{n}\in\mathbb{Z}^d} g(\frac{\mathbf{x}-2\pi\mathbf{n}}{\alpha}) \tag{5.8}$$

and

$$\omega_\alpha(\mathbf{x}) = \int_\Omega \omega(\mathbf{y})\,h_\alpha(\mathbf{x}-\mathbf{y})\,d\mathbf{y}. \tag{5.9}$$

Then $\omega_\alpha \in C^\infty(\mathbb{T}^d)$,

$$\min_{\mathbf{x}\in\mathbb{T}^d}\omega_\alpha(\mathbf{x}) \ge \min_{\mathbf{y}\in\mathbb{T}^d}\omega(\mathbf{y})\cdot\int_\Omega h_\alpha(\mathbf{y})\,d\mathbf{y} = \min_{\mathbf{y}\in\mathbb{T}^d}\omega(\mathbf{y}) > 0, \tag{5.10}$$

and

$$\|\omega_\alpha - \omega\|_{C^1(\mathbb{T}^d)} \le C \max_{\substack{\mathbf{x},\mathbf{y}\in\Omega \\ \operatorname{dist}(\mathbf{x}-\mathbf{y},(2\pi\mathbb{Z})^d)\le\alpha}} \big\{|\omega(\mathbf{x})-\omega(\mathbf{y})| + |\mathbf{D}\omega(\mathbf{x}) - \mathbf{D}\omega(\mathbf{y})|\big\}. \tag{5.11}$$

Note that

$$\|\mathbf{D}\big(\ln(\frac{\omega}{\omega_\alpha})\big)\|_{C(\mathbb{T}^d)} \le \frac{\|\omega-\omega_\alpha\|_{C(\mathbb{T}^d)}\|\omega\|_{C^1(\mathbb{T}^d)}}{(\min_{\mathbb{T}^d}\omega)^2} + \frac{\|\mathbf{D}\omega - \mathbf{D}\omega_\alpha\|_{C(\mathbb{T}^d)}}{\min_{\mathbb{T}^d}\omega}.$$

This, together with (5.11) and the fact that $\mathbf{D}\omega$ is uniformly continuous on $\mathbb{T}^d$, implies that, for any $\gamma \in (0,1)$,

$$\|\mathbf{D}\big(\ln(\frac{\omega}{\omega_\alpha})\big)\|_{C(\mathbb{T}^d)} < \gamma$$

if α is sufficiently small. Estimate (5.7) then follows by choosing $\widetilde{\omega} = \omega_\alpha$ with a sufficiently small α. The proof is complete.

References

[1] M.Sh. Birman and T.A. Suslina, *Two-dimensional periodic magnetic Hamiltonian is absolutely continuous*, Algebra i Analiz **9(1)** (1997), 32-48; English transl. in St. Petersburg Math. J. 9(1) (1998), 21-32.

[2] ———, *Absolute continuity of the two-dimensional periodic magnetic Hamilton ian with discontinuous vector-valued potential*, Algebra i Analiz **10(4)** (1998), 1-36; English transl. in St. Petersburg Math. J. 10(4) (1999).

[3] ———, *A periodic magnetic Hamiltonian with a variable metric. The problem of absolute continuity*, Algebra i Analiz **11(2)** (1999), 1-40; English transl. in St. Petersburg Math. J. 11(2) (2000).

[4] M.Sh. Birman, R.G. Shterenberg, and T.A. Suslina, *Absolute continuity of the spectrum of a two-dimensional Schrödinger operator with potential supported on a periodic system of curves*, ESI Preprint 934 (2000).

[5] L. Danilov, *On the spectrum of the Dirac operator with periodic potential in* $\mathbb{R}^n$, Theor. and Math. Phys. **85** (1990), 1039-1048.

[6] R. Hempel and I. Herbst, *Strong magnetic fields, Dirichlet boundaries, and spectral gaps*, Comm. Math. Phys. **164** (1995), 237-259.

[7] ———, *Bands and gaps for periodic magnetic Hamiltonians*, in Operator Theory: Advances and Applications **78** (1995), 175-184.

[8] T. Kato, *Perturbation Theory for Linear Operators*, Second edition, Springer-Verlag, 1979.

[9] C.E. Kenig, *Restriction theorems, Carleman estimates, uniform Sobolev inequalities and unique continuation*, Lecture Notes in Math. **1384** (1989), 69-90.

[10] P. Kuchment, *Floquet Theory for Partial Differential Equations*, Birkhäuser Verlag, 1993.

[11] ———, *Mathematics of photonic crystals*, to appear in Mathematical Modeling in Optical Science, SIAM (2001).

[12] P. Kuchment and S. Levendorskii, *On the structure of spectra of periodic elliptic operators*, Preprint (2000).

[13] A. Morame, *Absence of singular spectrum for a perturbation of a two-dimensional Laplace-Beltrami operator with periodic electro-magnetic potential*, J. Phys. **31** (1998), 7593-7601.

[14] M. Reed and B. Simon, *Methods of Modern Mathematical Physics*, vol. II and IV, Academic Press, 1978.

[15] Z. Shen, *On fundamental solutions of generalized Schrödinger operators*, J. Functional Analysis **167** (1999), 521-564.

[16] ———, *On absolute continuity of the periodic Schrödinger operators*, International Math. Res. Notices **2001(1)** (2001), 1-32.

[17] ———, *The periodic Schrödinger operator with potentials in the Morrey-Campanato class*, Preprint (1999).

[18] ———, *Absolute continuity of periodic Schrödinger operators with potentials in the Kato class*, to appear in Illinois J. Math..

[19] A.V. Sobolev, *Absolute continuity of the periodic magnetic Schrödinger operator*, Invent. Math **137(1)** (1999), 85-112.

[20] E.M. Stein, *Singular Integrals and Differentiability Properties of Functions*, Princeton Univ. Press, 1970.

[21] L.E. Thomas, *Time dependent approach to scattering from impurities in a crystal*, Comm. Math. Phys. **33** (1973), 335-343.

[22] C. Tolmasky, *Exponentially growing solutions for nonsmooth first-order perturbations of the Laplacian*, SIAM J. Math. Anal. **29** (1998), 116-133.

DEPARTMENT OF MATHEMATICS, UNIVERSITY OF KENTUCKY, LEXINGTON, KY 40506

E-mail address: shenz@ms.uky.edu

Contemporary Mathematics
Volume **277**, 2001

The use of Rellich identities on certain nongraph boundaries

Gregory C. Verchota

In memory of Eugene B. Fabes

Introduction

In 3 dimensions consider the open cone exterior to the union of two closed quarter-spaces $x \leq 0$, $z \geq 0$ and $y \leq 0$, $z \leq 0$ (the two bricks; see [**KW**]). The boundary of this cone or two-brick domain includes, in the $x - y$ plane, the 2nd and 4th quadrants. The outward directed normal vectors for these two surfaces point in precisely opposite directions while the two surfaces meet at the origin. Therefore the boundary of this cone cannot be represented locally at every point of the boundary as the graph of a Lipschitz function in some rectangular coordinate system; the cone is not a Lipschitz domain in the sense of [**D**] even though it is the union of two Lipschitz domains. More generally, bounded polyhedra are not Lipschitz domains.

Dahlberg's analysis of harmonic measure exploited extending the Green function across boundaries given as graphs. Jerison and Kenig [**JK**] introduced the Rellich-Necas-Payne-Weinberger formula into the study of harmonic functions in Lipschitz domains. The power of this formula (see line (2) below) depended crucially on the strict positivity of the inner product of the normal vector field with a smooth vector field transverse to the boundary ($N \cdot W$ in line (2)), i.e. again depended on the boundary as a graph.

Because this method of [**JK**] has proven so effective for a great many problems (see [**K**] for a survey) it is the intention of this paper to make it work nevertheless for polyhedra in 3 dimensions. In line (2) below, $N \cdot W$ is again bounded and bounded from below away from zero. The difficulty is transferred to the solid integral where derivatives of the vector field W appear that necessarily blow up as the origin (a vertex of the polyhedron) is approached. The proof of Lemma 3 resolves this problem.

As an application the invertibility of the classical layer potentials on polyhedra is then proven (Theorem 6). This last theorem was essentially proven for the polyhedra considered here by S. Rempel [**Re**] (see [**Ra**] and [**E**]) by using the Mellin transform to decompose the boundary operators into operators living on polygons

The author was supported in part by NSF Grant #9706648.

on spheres centered at each corner of a polyhedron. See "Concluding remarks" 2 and 4 below.

Also included here is a discussion of the classical layer potential techniques on what could be termed cones with Lipschitz facets. See "Layer potentials for bounded polyhedra" below.

Preliminaries

For a domain $D \subset \mathbf{R}^n$ and some α between 0 and 1 small enough define the nontangential approach region for each Q in the boundary ∂D of D by $\Gamma(Q) = \Gamma_\alpha(Q) = \{X \in D : \text{dist}\,(X, \partial D) > \alpha|X - Q|\}$. For any real or vector-valued function F defined in D define the nontangential maximal functions of F by

$$N(F)(Q) = \sup_{X\in\Gamma(Q)} |F(X)|.$$

When D is a bounded Lipschitz domain with connected boundary, [**JK**] (see also [**V**]) establishes the following Neumann and regularity theorems. Here and throughout this paper Lebesgue spaces on the boundary are taken with respect to surface measure $d\sigma$ (or dQ). L_1^p denotes Sobolev spaces of first derivatives.

THEOREM N (Jerison-Kenig). *Given $g \in L^2(\partial D)$ with $\int_{\partial D} g\, d\sigma = 0$ there exists a unique (up to additive constants) harmonic function u in D such that*

$$(i)\quad \lim_{\substack{X\to Q\\ X\in\Gamma(Q)}} N_Q \cdot \nabla u(X) = g(Q)\ a.e.\, d\sigma(Q)$$

$$(ii)\ N(\nabla u) \in L^2(\partial D).$$

Further $\|N(\nabla u)\|_{L^2(\partial D)} \leq C\|g\|_{L^2(\partial D)}$ with C depending only on the Lipschitz nature of D and $\lim_{\substack{X\to Q\\ X\in\Gamma(Q)}} \nabla u(X) := \nabla u(Q)$ exists a.e. $d\sigma(Q)$.

THEOREM R (Jerison-Kenig). *Given $g \in L_1^2(\partial D)$ there exists a unique harmonic function u in D such that*

$$(i)\quad \lim_{\substack{X\to Q\\ X\in\Gamma(Q)}} u(X) = g(Q)\ a.e.\, d\sigma(Q)$$

$$(ii)\ N(\nabla u) \in L^2(\partial D).$$

Further $\|N(\nabla u)\|_{L^2(\partial D)} \leq C\|g\|_{L_1^2(\partial D)}$ with C as in Theorem N and $\lim_{\substack{X\to Q\\ X\in\Gamma(Q)}} \nabla u(X) := \nabla u(Q)$ exists a.e. $d\sigma(Q)$ with the tangential components of the gradient $\nabla_T u(Q) = \nabla_T g(Q)$a.e. $d\sigma(Q)$.

Note that every solution of the Neuman problem is a solution to a regularity problem and vice versa.

Let ω denote a simply connected domain on the unit sphere $\mathbf{S}^2 \subset \mathbf{R}^3$ with complement $\omega^c \subset \mathbf{S}^2$ nonempty. Letting $\Omega_0 = \{X \in \mathbf{R}^3 : X|X|^{-1} \in \omega,\ 1 < |X| < 2\}$, ω is defined to be a Lipschitz domain if and only if $\Omega_0 \subset \mathbf{R}^3$ is a Lipschitz domain. In this case $\bar{\omega}^c$ is also a simply connected Lipschitz domain.

For a.e. $Q \in \partial\Omega_0$ let $N_Q = N$ denote the outer unit normal vector to Ω_0. When $\theta = Q|Q|^{-1} \in \partial\omega$, N_θ also denotes the outer unit normal to ω; when $Q|Q|^{-1} \in \omega$, N_Q points in the radial direction $\pm Q|Q|^{-1}$. Let $W : \mathbf{S}^2 \to \mathbf{R}^3$ be a C^∞ vector field that is transverse to $\partial\omega$, i.e. the scalar product $W_\theta \cdot N_\theta$ is bounded below by a positive constraint uniformly (a.e.) in $\theta \in \partial\omega$. The domain ω may be continuously deformed on $\mathbf{S}^2$ to a spherical cap so that the Lipschitz properties (on $\mathbf{S}^2$) of the resulting homotopy-family of domains are uniform. It is in addition possible to require of W that it remain uniformly transverse to the homotopy. In particular the constants of Theorems N and R will be uniform over the homotopy. The cone in $\mathbf{R}^3$ induced by the spherical cap is Lipschitz, unlike in general that induced by ω. Thus, if need be, the resulting homotopy of cones can be exploited in the way that will be mentioned below in "Invertibility of layer potentials."

Let $\Omega \subset \mathbf{R}^3$ denote the interior of a cone with vertex at the origin such that $\omega = \Omega \cap \mathbf{S}^2$ is a simply connected Lipschitz domain in $\mathbf{S}^2$. For integers j define $\Omega_j = \Omega \cap \{X : 2^j < |X| < 2^{j+1}\}$ and define Ω_j^* to be the interior of the closure of $\Omega_{j-1} \cup \Omega_j \cup \Omega_{j+1}$. Then Ω_j and Ω_j^* are scalings of the Lipschitz domains Ω_0 and Ω_0^* respectively.

The Rellich formula for conical points

A function will be said to be a *local* solution to the Neumann problem (or regularity problem) in Ω if it is a solution of Theorem N for a finite union

(1) $$D = \cup_{j_1 \le j \le j_2} \Omega_j^*.$$

LEMMA 1. *Let u be a local solution to the Neumann problem in Ω. Then for each j ($j_1 < j < j_2$)*

$$\int_{\Omega_j} |\nabla u(X)|^2 \frac{dX}{|X|} \le C\Big(\int_{\partial\Omega \cap \partial\Omega_j^*} \big(\frac{\partial u}{\partial N}\big)^2 \, d\sigma + \int_{\Omega_j^*} (X \cdot \nabla u(X))^2 \frac{dX}{|X|^3}\Big)$$

where C is finite and depends only on Ω and not on j.

PROOF. By rescaling it suffices to take $j = 0$. Let $\Omega_t = \Omega \cap \{X : t^{-1} < |X| < 2t\}$ for $1 < t < 2$. We have diam $\Omega_t < 8$ for all t. From Theorem N (and its proofs) there is a constant independent of t such that

$$\int_{\Omega_0} |\nabla u|^2 \frac{dX}{|X|} \le \int_{\Omega_t} |\nabla u|^2 \, dX \le C \int_{\partial\Omega_t} \big(\frac{\partial u}{\partial N}\big)^2 \, d\sigma.$$

For $X \in \partial\Omega_t \cap \Omega$, $|\frac{\partial u}{\partial N}(X)| = |X \cdot \nabla u(X)| \, |X|^{-1}$. Using this, integrating in t and using $|X| < 4$, the lemma follows for $j = 0$. □

Define $\Omega_- = \Omega \cap \{X : |X| < 1\}$ and $\Omega_-^* = \Omega \cap \{X : |X| < 2\}$.

LEMMA 2. *Let u be defined throughout Ω_-^* and a local solution to the Neumann problem in Ω. Then*

$$\int_{\Omega_-} |\nabla u(X)|^2 \frac{dX}{|X|} \le C\Big(\int_{\partial\Omega \cap \partial\Omega_-^*} \big(\frac{\partial u}{\partial N}\big)^2 \, d\sigma + \int_{\Omega_-^*} (X \cdot \nabla u(X))^2 \frac{dX}{|X|^3} + \int_{\Omega_0} u^2(X) \, dX\Big)$$

where C is as in Lemma 1.

PROOF. Using a smooth radial cut-off function that has its derivative supported in Ω_0, the lemma follows from Lemma 1 and monotone convergence. □

Let D be as in (1) and let W and N be as above. Extend W to $\mathbf{R}^3 \setminus \{0\}$ by $W(X) = W(X|X|^{-1})$. Then W is transverse to $\partial\Omega$ also and may be decomposed into normal and tangential parts, $W = (W \cdot N)N + W_T$. Denote by $\nabla_T u$ the tangential component of ∇u. For a solution of the Neumann problem in D the following well-known integration by parts formula (Rellich formula) holds:

$$(2)\quad \int_{\partial D} |\nabla_T u|^2 N\cdot W\,d\sigma = \int_{\partial D} \Big(\frac{\partial u}{\partial N}\Big)^2 N\cdot W\,d\sigma + 2\int_{\partial D} \frac{\partial u}{\partial N}\nabla u\cdot W_T\,d\sigma + \int_D \operatorname{div} W|\nabla u|^2 - 2D_iW_jD_iuD_ju\,dX$$

where summation convention $1\le i,j\le 3$ on the components of W and ∇u is used.

LEMMA 3. *Let u be harmonic in Ω_-^* with $N(\nabla u)$ square integrable on $\partial\Omega_-^*$. Then the integrals of Lemma 2 are finite and*

$$(3)\quad \int_{\partial\Omega_-\cap\partial\Omega} |\nabla u|^2\,d\sigma \le C_1\Big(\int_{\partial\Omega_-^*\cap\partial\Omega}\Big(\frac{\partial u}{\partial N}\Big)^2 d\sigma + \int_{\Omega_0^*}|\nabla u|^2\,dX\Big)$$

where $C_1<\infty$ depends only on the Lipschitz nature of ω.

PROOF. The first assertion follows easily since $|X|^{-1}\,dX$ is a Carleson measure for Ω. Since the last integral of (2) is dominated by a constant C_W depending on the derivative of W times the left side of Lemma 2, the domains D of (2) may be replaced with Ω_- by a limiting argument using also the nontangential control of the gradient. Now the last integral of (2) is dominated by C_W times the right side of Lemma 2.

Let γ be a real number. By adding the vector field $\gamma|X|^{-1}X$ to W, the left side of (2) and the first integral on the right of (2) are unchanged except for the interior surface $\partial\Omega_-\cap\Omega$ since N and X are orthogonal. The last integral yields precisely the new term $2\gamma\int_{\Omega_-}(X\cdot\nabla u)^2\frac{dX}{|X|^3}$. By choosing γ so that $2\gamma<-CC_W$, with C as in Lemma 2, (2) now implies

$$(4)\quad \int_{\partial\Omega_-\cap\partial\Omega}|\nabla_T u|^2\,d\sigma \le C_1\Big(\int_{\partial\Omega_-^*}\Big(\frac{\partial u}{\partial N}\Big)^2 d\sigma + \int_{\Omega_0}u^2\,dX\Big) + C_\gamma\Big(\int_{\partial\Omega_-}\Big|\frac{\partial u}{\partial N}\Big|\,|\nabla_T u|\,d\sigma + \int_{\partial\Omega_-\cap\Omega}|\nabla u|^2\,d\sigma\Big)$$

where C_1 depends on the Lipschitz nature of ω and C_γ depends on γ.

The second to last integral of (4) may now be controlled through Young's inequality. Since any constant can be subtracted from u, Poincaré's inequality applies to the second integral. By averaging scalings of Ω_-, the boundary integrals of (4) that are interior to Ω may be controlled by the last integral of (3). □

Lemma 3 states the main inequality needed to solve the Neumann problem in the (non-Lipschitz) cones Ω. The next lemma serves to solve the regularity problem. To prove it Lemma 1 must be restated with the tangential gradient replacing the normal derivative. To establish the restatement Brown's inequality for the mixed (Neumann and regularity) problem may be used. We state it for the domains Ω_0 (which satisfy Brown's geometric conditions) after an application of Poincaré's inequality on $\partial\Omega_0\cap\partial\Omega$:

THEOREM M (Brown [**B**], Theorem 2.1). *Let u be a solution of Theorem N in Ω_0. Then there is a constant $C<\infty$ depending only on the Lipschitz nature of Ω_0*

such that

$$\int_{\partial\Omega_0} N(\nabla u)^2\, d\sigma \le C\Big(\int_{\partial\Omega_0\cap\partial\Omega} |\nabla_T u|^2\, d\sigma + \int_{\partial\Omega_0\cap\Omega} (\frac{\partial u}{\partial N})^2\, d\sigma\Big).$$

A restated Lemma 2 now follows. The Rellich formula (2) may be obviously rearranged to serve to estimate the normal derivative instead of the tangential gradient and the proof of Lemma 3 follows with γ chosen so $2\gamma > CC_W$.

LEMMA 4. *With the same hypothesis as in Lemma 3 and with C_1 as in Lemma 3,*

$$\int_{\partial\Omega_-\cap\partial\Omega} |\nabla u|^2\, d\sigma \le C_1\Big(\int_{\partial\Omega_-^*\cap\partial\Omega} |\nabla_T u|^2\, d\sigma + \int_{\Omega_0^*} |\nabla u|^2\, dX\Big).$$

Bounded Polyhedra

Let $D \subset \mathbf{R}^3$ be a simply connected domain so that $\bar{D}^c$ is simply connected. Suppose further that given any point of the boundary all open balls centered there with small enough radii have an intersection with D that is simply connected. In addition let D be the interior of the union of finitely many polytopes (bounded convex polyhedra). D is in general a nonconvex bounded polyhedron. The boundary of D contains at most finitely many points for each of which the *local* boundary of D cannot be realized as the graph of a function in some rectangular coordinate system. However, by translation and rescaling if necessary, the intersection of D with the unit sphere will be equal to a spherical polygon ω with geodesic sides and thus be in the class of spherical Lipschitz domains described above. Furthermore the corresponding truncated cone Ω_-^* can be made to equal $D \cap \{X : |X| < 2\}$. Therefore the boundary estimates of Lemmas 3 and 4 hold for D around these points, while the corresponding estimates of [**JK**] hold elsewhere on the boundary. Consequently inequalities like

$$\int_{\partial D} |\nabla u|^2\, d\sigma \le C\Big(\int_{\partial D} (\frac{\partial u}{\partial N})^2\, d\sigma + \int_D |\nabla u|^2\, dX\Big)$$

follow where C depends only on any Lipschitz constants associated to D in the ways described above. The last integral may now be eliminated by the divergence theorem and Poincaré's inequality in the standard ways since ∂D is connected.

THEOREM 1. *Let D be a bounded polyhedron of $\mathbf{R}^3$ as described above. Let u be harmonic in D with $N(\nabla u)$ square integrable on the boundary. Then*

$$(i) \int_{\partial D} |\nabla u|^2\, d\sigma \le C \int_{\partial D} (\frac{\partial u}{\partial N})^2\, d\sigma \text{ and}$$

$$(ii) \int_{\partial D} |\nabla u|^2\, d\sigma \le C \int_{\partial D} |\nabla_T u|^2\, d\sigma$$

For R large enough, the domains $\{X : |X| < R\} \setminus \bar{D}$ also admit to the above analysis except for the last use of Poincaré's inequality since the boundary is not connected. Standard limiting arguments establish the next theorem for the exteriors of bounded polyhedra.

THEOREM 2. *Let D be a bounded polyhedron of $\mathbf{R}^3$. Let u be harmonic in $\bar{D}^c$ vanishing at infinity with $N(\nabla u)$ square integrable on the boundary. Then the inequalities of Theorem 1 continue to hold as long as $C(\int_{\partial D} u\, d\sigma)^2$ is added to the right sides.*

Layer potentials for bounded polyhedra

Consider now the classical single layer potential for the polyhedral domains D described above and $f \in L^p(\partial D)$ $(1 < p < \infty)$

$$(5) \qquad Sf(X) = (4\pi)^{-1} \int_{\partial D} |X - Q|^{-1} f(Q)\, dQ \quad (X \in \mathbf{R}^3 \setminus \partial D).$$

In order for the previous lemmas and theorems to apply to Sf and its gradient and in order to justify any subsequent use of the Gauss divergence theorem or other integration formulas in D, it is required that the gradient of this potential have nontangential maximal function in L^p. Consider, more generally, a spherical Lipschitz domain ω, as described above, representing the intersection of a conical corner of a domain D with the unit sphere. Denote this conical corner by $\partial\Omega_-$.

We now give an argument establishing what we need for the nontangential maximal function on these kind of conical corners. (Alternatively the reader can refer to pp. 11–14 of the book [**DS**] where by Proposition 1.28 the maximal singular integral operator $\sup_{\varepsilon > 0} \Big| \int_{\substack{\partial\Omega_- \\ |P-Q|>\varepsilon}} \frac{P - Q}{|P - Q|^3} f(Q)\, dQ \Big|$, $(P \in \partial\Omega_-)$ associated with the gradient of (5) is shown to be bounded for $1 < P < \infty$ since $\partial\Omega_-$ satisifies the "big pieces of Lipschitz graphs" condition (Def. 1.26) of that monograph. By the techniques of [**FJR**] one can then pass to the nontangential maximal functions.)

By definition of Lipschitz domain it will suffice to study (5) when f is supported in one of a finite number of pieces (Lipschitz facets) of $\partial\Omega_-$ produced by crossing a piece of $\partial\omega$ with distances to the origin $\rho\,(0 < \rho < 1)$. Specifically, there exists M and $a > 0$ small enough so that standard spherical coordinates $\theta(-\pi < \theta \leq \pi)$ and $\phi(0 \leq \phi \leq \pi)$ can be introduced at any point of $\partial\omega$ making that point $(0, \frac{\pi}{2})$ and so that $\{(\theta, \phi) : |\theta| < 100a, |\phi| < 100Ma\} \cap \omega = \{(\theta, \phi) : |\theta| < 100a, \phi(\theta) < \phi < 100Ma\}$ where $|\phi'(\theta)| \leq M$ (a.e. θ). Further $a > 0$ may be chosen small so that $\phi(\theta)$ may be extended to $|\theta| < \frac{\pi}{2}$ in such a way that $\phi(\theta) - \frac{\pi}{2}$ is supported in $|\theta| < 200a$, $|\phi'(\theta)| \leq M$, and $z = \rho\cos(\phi(\theta))\,(|\theta| \leq \frac{\pi}{2})$, $z = 0\,(\frac{\pi}{2} < |\theta| < \pi)\,(\rho > 0)$ defines a Lipschitz function of the standard rectangular coordinates (x, y). For any $b < 100a$, $\{(\rho, \theta, \phi(\theta)) : |\theta| < b \text{ and } 0 < \rho < 1\}$ will be called a *Lipschitz facet of radius* b of the truncated cone Ω_- and the graph of z above will be called its supporting Lipschitz plane. A finite number of Lipschitz facets of radius a will cover $\partial\Omega_-$. If f in (5) is supported in one of them, then Sf is also a potential defined on its supporting Lipschitz plane. Therefore ∇Sf has nontangential maximal function in L^p of the corresponding Lipschitz facet of radius $100a$ by [**CMM**] (see also [**C**], [**FJR**], [**V**]).

Hence it suffices to consider the nontangential maximal function of ∇Sf on a facet $\mathcal{B}$ of radius a that is sufficiently distant on the unit sphere from the supporting facet $\mathcal{A}$ of f. Sf and its gradient are continuous functions on $\mathcal{B}$. Moreover, using a device of [**B**], the surface measure of $\mathcal{B}$ is a Carleson measure for the Lipschitz domain above (or below or both) the supporting Lipschitz plane of $\mathcal{A}$. Consequently ∇Sf on $\mathcal{B}$ is bounded in L^p by the nontangential maximal function of ∇Sf (e.g. [**S**], p. 236) for the supporting plane of $\mathcal{A}$ which in turn by [**CMM**] is bounded by the L^p norm of f on $\mathcal{A}$.

As for the nontangential maximal function fix $P \in \mathcal{B}$ and let $X \in \Gamma_\alpha(P)$. The operator kernel of ∇Sf satisfies what are called the standard estimates (e.g. [**J**], p. 47 or [**DS**], Def. 1.20). Consequently for $|X - P|$ small (depending on

$100a|P|)$, $\left|\frac{X-Q}{|X-Q|^3} - \frac{P-Q}{|P-Q|^3}\right| \le \frac{C|X-P|}{|P-Q|^3} \le \frac{C|P|}{|P-Q|^3}$ for all Q in $\mathcal{A}$, the support of f. Using this, $\nabla Sf(X) - \nabla Sf(P)$ is controlled in the standard way (see [**FJR**]) by the Hardy-Littlewood maximal function of f at any $\widetilde{P} \in \mathcal{A}$, $|\widetilde{P}| = |P|$. For large $|X-P|$, $\nabla Sf(X)$ may be controlled by Hardy-Littlewood, again at $|\widetilde{P}|$ as above, together with truncated operators $\int_{\substack{\mathcal{A} \\ |Q|>t}} \frac{P-Q}{|P-Q|^3} f(Q)dQ$ where it suffices to always take $t > 2|P|$. But then P may be replaced with a $|\widetilde{P}|$ modulo Hardy-Littlewood and the truncated operator controlled by the maximal singular integral operator on the supporting Lipschitz plane of $\mathcal{A}$. Using these pointwise estimates, $N(\nabla Sf)$ on $\mathcal{B}$ is controlled in L^p by ∇Sf on $\mathcal{B}$ and by various maximal operators on $\mathcal{A}$ known to also be controlled by f.

The same applies to the double layer potential

$$\mathcal{K}f(x) = (4\pi)^{-1} \int_{\partial D} \frac{N_Q \cdot (Q-X)}{|Q-X|^3} f(Q)dQ \; (X \in \mathbf{R}^3 \setminus \partial D) \tag{6}$$

These arguments yield the following theorem as a special case.

THEOREM 3. *Let D be a bounded polyhedron of $\mathbf{R}^3$, $1 < p < \infty$, and let $f \in L^p(\partial D)$. Then the nontangential maximal functions formed either in D or in $\bar{D}^c$ of ∇Sf and $\mathcal{K}f$ are bounded in $L^p(\partial D)$ by the L^p-norm of f times a finite constant that depends only on the various Lipschitz constants used above to describe D.*

Theorem 3 implies by standard arguments (see [**V**]) that (6) has a nontangential trace a.e. on the boundary equal to $\frac{1}{2}f + Kf$ when X approaches from inside D ($-\frac{1}{2}f + Kf$ from outside) where $K : L^p(\partial D) \to L^p(\partial D)$ $(1 < p < \infty)$ is a bounded Calderón-Zygmund singular integral (e.g. [**J**]). Let I denote the identity operator. The adjoint operator $\frac{1}{2}I + K^*$ is also Calderón-Zygmund and represents the normal derivative of (5) from the outside; more precisely

$$\begin{aligned} &\lim_{X \to P} N_P \cdot \nabla Sf(X) = (\tfrac{1}{2}I + K^*)f(P) \text{ a.e. } (X \in \bar{D}^c) \\ &\qquad\qquad\text{and} \\ &\lim_{X \to P} N_P \cdot \nabla Sf(X) = (-\tfrac{1}{2}I + K^*)f(P) \text{ a.e. } (X \in D) \end{aligned} \tag{7}$$

where the limits are taken in nontangential cones at each point $P \in \partial D$. In addition the inside and outside limits of Sf and its *tangential* derivatives agree on the boundary.

The arguments of [**V**] apply here so that Theorems 1, 2 and 3 together with formulas (7) yield the following L^2 result.

THEOREM 4. *Let D be a bounded polyhedron in $\mathbf{R}^3$. Then there is a constant C as in Theorem 3 so that*

$$\text{(i)} \int_{\partial D} ((\tfrac{1}{2}I + K^*)f)^2 \, d\sigma \le C\Big(\int_{\partial D} ((\tfrac{1}{2}I - K^*)f)^2 \, d\sigma + \big(\int_{\partial D} Sf \, d\sigma\big)^2\Big)$$

and

$$\text{(ii)} \int_{\partial D} ((\tfrac{1}{2}I - K^*)f)^2 \, d\sigma \le C\Big(\int_{\partial D} ((\tfrac{1}{2}I + K^*)f)^2 \, d\sigma + \big(\int_{\partial D} Sf \, d\sigma\big)^2\Big).$$

Invertibility of boundary layer potentials

Let $L^p_0(\partial D)$ denote the codimension one subspace of L^p-functions with integral zero over the boundary. Standard arguments ([**FJR**], also [**V**]) show that $\frac{1}{2}I+K^*: L^2(\partial D)\to L^2(\partial D)$ is one-to-one while $-\frac{1}{2}I+K^*: L^2_0(\partial D)\to L^2_0(\partial D)$ is one-to-one. As in [**V**] Theorem 4 implies that each of $\frac{1}{2}I\pm K^*$ has closed range in L^2.

Proving that the layer potential operators of (7) are onto L^2 and L^2_0 respectively may be done in a couple different ways. However, using surjectivity on a sequence of approximating smooth domains as in [**V**] does not appear possible given the nature of polyhedral corners. One solution is to use a method of continuity argument (e.g. [**GT**], p. 75) exploiting the uniform constants of the theorems and lemmas here over the above mentioned homotopy of cones determined by the spherical domains ω and a circular spherical cap. Cones formed by the latter, of course, have boundaries in $\mathbf{R}^3$ given as Lipschitz graphs which means that a member of the homotopy family of operators (7) would be known to be onto. The operator homotopy, unlike [**GT**], would be formed by introducing into the operator kernels a convex combination of the spherical Lipschitz functions describing the boundaries of ω and the spherical cap. (The author thanks Prof. A. McIntosh for this insight.) Unlike the approximation in [**V**] this allows, for small perturbations of the homotopy parameter, the Calderón-Zygmund operators to be close in *operator* norm. The uniform bound from below is guaranteed by Theorem 4 which could have been stated for domains with Lipschitz facets.

A more functional analytic method is due to M. Mitrea [**M**]. For $|\lambda|$ large, $\lambda I+K^*$ is an isomorphism while, as shown above, $\pm\frac{1}{2}I+K^*$ has a well-defined index (is semi-Fredholm). By a continuity method the index of $\lambda I+K^*$ will be preserved as λ approaches $\pm\frac{1}{2}$ through the reals, if the bound from below

$$\text{(8)}\qquad \int_{\partial D} f^2\,d\sigma\le C\int_{\partial D}\left((\lambda+K^*)f\right)^2 d\sigma$$

can be made uniform in λ and f in L^2 or L^2_0. Theorem 4 proves (8) for $\lambda=\pm\frac{1}{2}$ as in [**V**] (pp. 585–586) and therefore for all λ close enough to $\pm\frac{1}{2}$. To prove (8) for all λ in some large enough compact interval of either positive or negative real numbers, the Rellich formula (2) is exploited as in [**EFV**]. By following the instructions in Proposition 4.3 of [**MT**] with $u=Sf$, (2) with $D=\Omega_-$ together with $D=\bar\Omega^c\cap\{|X|<1\}$ yields

$$\begin{aligned}\text{(9)}\quad (\lambda^2-\tfrac{1}{4})&\int_{\partial\Omega_-\cap\partial\Omega} f^2N\cdot W\,d\sigma+\int_{\partial\Omega_-\cap\partial\Omega}|\nabla_T u|^2N\cdot W\,d\sigma\\ &=\int_{\partial\Omega_-\cap\partial\Omega}\left((\lambda+K^*)f\right)^2N\cdot W+2(\lambda+K^*)f\nabla u\cdot W_T\,d\sigma\\ &\quad-(\lambda+\frac{1}{2})\int_{\bar\Omega^c\cap\{|X|<1\}}\mathrm{div}W|\nabla u|^2-2D_iW_jD_iuD_ju\,dX\\ &\quad-(\lambda-\frac{1}{2})\int_{\Omega_-}\mathrm{div}W|\nabla u|^2-2D_iW_jD_iuD_ju\,dX\\ &\quad+\int_{|Q|=1}\nabla u\text{ quadratic terms }d\sigma.\end{aligned}$$

In order to control the solid integrals a substitute for Lemma 1 is needed.

LEMMA 5. *Let D be a bounded polyhedron in $\mathbf{R}^3$ and $u = Sf$ as in* (5). *Let the cone Ω as in Lemma 1 satisfy $\partial D \cap \{|X| < 8\} = \partial\Omega \cap \{|X| < 8\}$. Let $|\lambda| \geq \frac{1}{2}$. Then for every $\delta\,(0 < \delta < 1)$ there is a finite constant $C_{\lambda,\delta}$ otherwise only depending on D such that*

$$|\lambda - \tfrac{1}{2}| \int_{\Omega_0} |\nabla u|^2 \frac{dX}{|X|} + |\lambda + \tfrac{1}{2}| \int_{\bar{\Omega}^c \cap \{1<|X|<2\}} |\nabla u|^2 \frac{dX}{|X|}$$
$$\leq \delta \Big(\int_{\partial\Omega_0^* \cap \partial\Omega} |\nabla_T u|^2 \, d\sigma + \int_{\frac{1}{2}<|X|<4} |\nabla u|^2 \frac{dX}{|X|} \Big)$$
$$+ C_{\lambda,\delta} \Big(\int_{\partial\Omega_0^* \cap \partial\Omega} ((\lambda + K^*)f)^2 \, d\sigma + \int_{\frac{1}{2}<|X|<4} (\nabla u \cdot X)^2 \frac{dX}{|X|^3} \Big).$$

PROOF. An application of the Gauss divergence theorem together with (7) (N is always outer to D) yields

$$(\lambda - \tfrac{1}{2}) \int_{\Omega_0} |\nabla u|^2 \, dX + (\lambda + \frac{1}{2}) \int_{\bar{\Omega}^c \cap \{1<|X|<2\}} |\nabla u|^2 \, dX$$
$$= (\lambda - \tfrac{1}{2}) \int_{\partial\{1<|X|<2\} \cap \Omega} u \frac{\partial u}{\partial N} \, d\sigma + (\lambda + \tfrac{1}{2}) \int_{\partial\{1<|X|<2\} \cap \bar{\Omega}^c} u \frac{\partial u}{\partial N} \, d\sigma$$
$$- \int_{\partial\Omega_0 \cap \partial\Omega} u(\lambda + K^*) f \, d\sigma.$$

The same can be done for the domains Ω_t, Young's inequality applied, the Poincaré inequality justified, and averaging used to establish the result. □

Lemma 5 rescales for the domains Ω_j of Lemma 1. *When $|\lambda|$ is strictly greater than $\frac{1}{2}$*, δ may be chosen small in order to hide the first solid integral on the right of Lemma 5 and the analogy to Lemma 2 becomes

$$|\lambda - \tfrac{1}{2}| \int_{\Omega_-} |\nabla u|^2 \frac{dX}{|X|} + |\lambda + \tfrac{1}{2}| \int_{\bar{\Omega}^c \cap \{|X|<1\}} |\nabla u|^2 \frac{dX}{|X|}$$
$$\leq 2\delta \int_{\partial\Omega_-^* \cap \partial\Omega} |\nabla_T u|^2 \, d\sigma + C_{\lambda,\delta} \Big(\int_{\partial\Omega_-^* \cap \partial\Omega} ((\lambda + K^*)f)^2 \, d\sigma$$
$$+ \int_{|X|<2} (\nabla u \cdot X)^2 \frac{dX}{|X|^3} + \int_{\mathbf{R}^3} |\nabla u|^2 \, dX \Big).$$

Now for each δ small enough and fixed the main point of this paper, choosing γ in the proof of Lemma 3, transforms (9) into

(10)

$$(\lambda^2 - \tfrac{1}{4}) \int_{\partial\Omega_- \cap \partial\Omega} f^2 N \cdot W \, d\sigma + \tfrac{1}{2} \int_{\partial\Omega_- \cap \partial\Omega} |\nabla_T u|^2 N \cdot W \, d\sigma$$
$$\leq \delta C_W \int_{\partial\Omega_-^* \cap \partial\Omega} |\nabla_T u|^2 \, d\sigma + C_{\lambda,\delta} \Big(\int_{\partial\Omega_-^* \cap \partial\Omega} ((\lambda + K^*)f)^2 \, d\sigma + \int_{\mathbf{R}^3} |\nabla u|^2 \, dX \Big)$$

where C_W in particular is independent of the choice of γ.

Using (10) a finite number of times around ∂D, taking care of the solid integral with a formula like the one used in the proof of Lemma 5 together with Poincaré, and choosing δ small establishes the following theorem.

THEOREM 5. *Let D be a bounded polyhedron in $\mathbf{R}^3$. Let $|\lambda| > \frac{1}{2}$. Then there is a finite constant as in Theorem 3 that also depends on λ so that for all square integrable f on the boundary of D*

$$\int_{\partial D} f^2 \, d\sigma \leq C_\lambda \int_{\partial D} \left((\lambda + K^*)f\right)^2 d\sigma.$$

As mentioned above such an inequality is true for λ in a neighborhood of $\frac{1}{2}$ and in a neighborhood of $-\frac{1}{2}$ if f has mean value zero.

THEOREM 6. *Let D be a bounded polyhedron of $\mathbf{R}^3$. Then there is an $\varepsilon > 0$ depending on the Lipschitz constants associated with D so that for all $p\,(2 - \varepsilon < p < 2 + \varepsilon)$ the classical boundary layer potentials $\frac{1}{2}I + K^*$ and $-\frac{1}{2}I + K^*$ are isomorphisms on $L^p(\partial D)$ and $L^p_0(\partial D)$ respectively.*

PROOF. The case $p = 2$ is established by the above results from Theorem 4 through Theorem 5. The perturbation from 2 may be done as in [**C2**]. □

Theorem 6 establishes existence for the Neumann problem for the interior and exterior of a bounded polyhedron. Uniqueness for $2 - \varepsilon < p < 2 + \varepsilon$ follows by integrating the square of the gradient as usual and using the nontangential control and Sobolev embedding to justify dominated convergence at the boundary.

Concluding remarks

1. If D satisfies the definition of bounded Lipschitz domain except for a *finite* number of points around which the boundary is given as a cone with *Lipschitz facets* as described above between lines (5) and (6), then the above discussion and proofs show that the theorems here continue to hold. This is an advantage of the Rellich formula approach over the Mellin transform approach and provides a simple example of solving the Neumann problem on the BPLG's of [**DS**].
2. Lemmas 1 and 2 would not be needed in higher dimensions. This is because the solid integrand of line (2) becomes $(n-3)|X|^{-1}\,|\nabla u|^2 + 2|X|^{-3}(X\cdot\nabla u)^2$ when W is replaced with $|X|^{-1}X$ in n dimensions. Thus the solid term can be made to have either sign by choosing γ. This indicates that higher dimensional non-Lipschitz polyhedra can be understood by the methods of this paper. The advantage of the Mellin approach in $\mathbf{R}^3$ is that operators of the form $I + T$ are obtained on the sphere with the spectral radius of T less than 1. Whether or not this device can also be exploited or easily expressed in higher dimensions is not clear to the author. Two-brick corners begin to appear on $\mathbf{S}^{n-1}$ for $n \geq 4$.
3. Given Theorem 6 there is no apparent impediment to proving a Hardy space $H^1(\partial D)$ result for the Neumann problem in bounded polyhedra of $\mathbf{R}^3$ by using the argument of [**DK2**]. The higher dimensional result becomes interesting, however. The use of Sobolev imbedding in [**DK2**] does not work in higher dimensions. On the other hand, the alternative method in [**DK**] relies on reflection across boundary that is given as a graph, a problem for the boundaries considered here.
4. Invertibility of the operator $\frac{1}{2}I + K$ in $L^p\,(2 - \varepsilon < p < \infty)$ and BMO would follow for the domains considered here, given the Hardy space result of the last remark. This points to the main advantage of the Mellin approach in 3

dimensions. For bounded polyhedra there, [**Ra**] established the invertibility of the double layer on the space of *continuous functions*. Thus classical solutions to the Dirichlet problem are shown to be represented by the double layer potential. This latter isomorphism is not readily obtainable by the methods here. On the other hand, it seems that the Mellin method for the double layer in higher dimensions will run into the problem mentioned above in remark 2.

5. Kenig and Pipher construct a domain in 3 dimensions with the property that for a $p > 2$ the Neumann problem fails while the Dirichlet problem holds for the conjugate exponent q, $q^{-1} + p^{-1} = 1$. See Section 2.9 of [**K**]. This domain turns out to be a two-brick type domain. To see this draw unit discs centered at $(-2, 0)$ and $(2, 0)$ in the plane. Connect them with a "bar" of width $2w$ symmetric about the x-axis. Call this ω. By drawing ω on a sphere and using it to produce a cone and then truncating the cone away from the origin one obtains the bounded domain Ω. Now, back to the plane, draw a line through the origin tangent to each disc. On the sphere this line represents a plane that is tangent to the cone along two distinct rays from the origin of $\mathbf{R}^3$. The two opposite pointing normals to this plane each serve as outer normals to the cone.

   In 2.9.5 of [**K**] for any $p > 2$ a harmonic function is constructed in an Ω with gradient nontangentially in L^2 but not L^p by choosing w small enough. Thus if there were another solution for this same good data but *nontangentially in* L^p then the uniqueness for the L^2 Neumann problem mentioned after Theorem 6 would be contradicted. Thus existence fails for p.

   In 2.9.6 of [**K**] it is shown that a $q_0 < 2$ exists so that for all $q > q_0$ the L^q-Dirichlet problem holds independently of $w > 0$. This means that the nontangential maximal function of any classical solution (continuous in $\bar{\Omega}$) is bounded in L^q by the L^q norm of its boundary values times a constant independent of w and the classical solution. An easy way to see this is to add the line segment from $(-2, 1)$ to $(2, 1)$ to the picture of ω and to define ω_+ to be the interior to the outer most lines and arcs. Similarly define ω_- by drawing the segment from $(-2, -1)$ to $(2, -1)$ so that $\omega = \omega_+ \cap \omega_-$. But $\omega_\pm$, put on the sphere, each produce cones and Lipschitz domains $\Omega_\pm$ with Lipschitz constants *independent* of w. Now estimates on the nontangential maximal functions can be reduced to considering only positive solutions in the non-Lipschitz domains Ω and comparing them via the classical maximum principle with positive solutions in the Lipschitz domains $\Omega_\pm$.

References

[B] R. M. Brown, *The mixed problem for Laplace's equation in a class of Lipschitz domains*, Comm. Part. Diff. Eq. **19** (1994), no. 7-8, 1217–1233.

[C] A. P. Calderón, *Cauchy integrals on Lipschitz curves and related operators*, Proc. Nat. Acad. Sci. U.S.A. **74** (1977), 1324–1327.

[C2] A. P. Calderón, *Boundary value problems for the Laplace equation in Lipschitz domains*, Recent Progress in Fourier Analysis (I. Peral and J.-L. Rubio de Francia, ed.), North-Holland, 1985.

[CMM] R. R. Coifman, A. McIntosh and Y. Meyer, *L'integrale de Cauchy definit un operateur borne sur L^2 pour les corbes lipschitzienne*, Ann. Of Math. **116** (1982), 361–387.

[D] B. E. J. Dahlberg, *On estimates of harmonic measure*, Arch Rational Mech. Anal. **65** (1977), 272–288.

[DK] B. E. J. Dahlberg and C. E. Kenig, *Hardy spaces and the Neumann problem in L^p for Laplace's equation in Lipschitz domains*, Ann. Of Math. **125** (1987), 437–465.

[DK2] B. E. J. Dahlberg and C. E. Kenig, *L^p estimates for the 3-dimensional systems of elastostatics on Lipschitz domains*, Analysis and partial differential equations, 621-634, Lecture Notes in Pure and Appl. Math. **122** Dekker, New York, 1990.

[DS] G. David and S. Semmes, *Analysis of and on Uniformly Rectifiable Sets*, A.M.S., Providence, RI, 1991.

[E] J. Elschner, *The double layer potential operator over polyhedral domains I: solvability in weighted Sobolev spaces*, Appl. Anal. **45** (1992), 117–134.

[EFV] L. Escauriaza, E. B. Fabes and G. C. Verchota, *On a regularity theorem for weak solutions to transmission problems with internal Lipschitz boundaries*, Proc. AMS **115** no. 4 (1992), 1069–1076.

[FJR] E. B. Fabes, M. Jodeit and N. M. Riviére, *Potential techniques for boundary value problems on C^1 domains*, Acta Math. **141** (1978), 165–186.

[GT] D. Gilbarg and N. S. Trudinger, *Elliptic partial differential equations of second order*, Springer-Verlag, Berlin, 1983.

[JK] D. S. Jerison and C. E. Kenig, *Boundary value problems on Lipschitz domains*, MAA Studies in Mathematics **23** Studies in Partial Differential Equations, W. Littman, ed. (1982), 1–68.

[J] J-L. Journé, *Calderón-Zygmund operators, pseudo-diferential operators and the Cauchy integral of Calderón*, Lecture Notes in Math., Springer-Verlag, Berlin, 1983.

[K] C. E. Kenig, *Harmonic analysis techniques for second order elliptic boundary value problems*, Providence, R.I., A.M.S., 1994.

[KW] J. Král and W. Wendland, *Some examples concerning applicability of the Fredholm-Radon method in potential theory*, Aplikace Matematiky **31** (1986), 293–308.

[M] M. Mitrea, *The method of layer potentials in electromagnetic scattering theory on nonsmooth domains*, Duke Math. J. **77** (1995), 111–133.

[MT] M. Mitrea and M. Taylor, *Boundary layer methods for Lipschitz domains in Riemannian manifolds*, J. Funct. Anal. **163** no. 2 (1999), 181–251.

[Ra] A. Rathsfeld, *The invertibility of the double layer potential operator in the space of continuous functions defined over a polyhedron*, Appl. Anal. **45** (1992), 135–177. A. Rathsfeld, *Erratum*, Appl. Anal. **56** (1995), 109-115.

[Re] S. Rempel, *Corner singularity for transmission problems in three dimensions*, Int. Eq. And Oper. Theory **12** (1989), 835–854.

[S] E. M. Stein, *Singular integrals and differentiability properties of functions*, Princeton University Press, Princeton, N.J., 1970.

[V] G. C. Verchota, *Layer potentials and regularity for the Dirichlet problem for Laplace's equation in Lipschitz domains*, J. Funct. Anal. **59** no. 3 (1984), 572–611.

DEPARTMENT OF MATHEMATICS, SYRACUSE UNIVERSITY, SYRACUSE, NEW YORK 13224
E-mail address: gverchota@syr.edu

Contemporary Mathematics
Volume **277**, 2001

L^2 boundedness of the Cauchy Integral and Menger curvature

Joan Verdera

ABSTRACT. In this paper we explain the relevance of Menger curvature in understanding the L^2 boundedness properties of the Cauchy Integral Operator. After introducing Menger curvature and describing its basic properties we proceed to prove the Coifman-McIntosh-Meyer Theorem on the Cauchy Integral on a Lipschitz graph. From this circle of ideas comes a new simple approach to the L^2 boundedness of the first Calderón commutator. We point out that the L^2 boundedness of the Cauchy Integral on a Lipschitz graph can be easily reduced to the boundedness of the first commutator. In the last section we describe the various steps in the solution of Vitushkin's conjecture on analytic capacity paying special attention to the role played by Menger curvature.

1. Introduction

The Menger curvature associated to a triple z_1, z_2 and z_3 of distinct points in the plane is the inverse of the radius of the circle passing through the given points. The Cauchy kernel $1/z$ and Menger curvature are related by a remarkable identity that has emerged recently and that has been shown to be extremely useful in dealing with L^2 boundedness of the Cauchy Integral Operator and, consequently, with analytic capacity. New striking results have been proved and new light has been shed on some classical theorems. The purpose of this article is to describe some of the main ideas behind these developments and try to make them accessible to analysts not necessarily familiar with the subject. On the other hand, we do not aim at any kind of completeness.

One intriguing fact about Menger curvature is that, up to now, no useful higher dimensional analogue that relates well to the Riesz kernels has been found. Therefore, for the Riesz transforms in higher dimensions no geometric approach is known that parallels the route available for the Cauchy kernel that will be described below. In fact, many higher dimensional versions of the results that will be presented in the next sections are still open.

2. Menger curvature

During the last few years the Menger's name has occurred in several parts of classical analysis. Some biographic notes will help place the person into a historical perspective.

2000 *Mathematics Subject Classification.* Primary 30–02, 30E20, 42–02, 42B20.

Karl Menger was born in Vienna in 1902 and died in Chicago in 1985. He wrote a Thesis on the definition of dimension for separable metric spaces under the direction of H. Hahn and became soon a mathematician of broad scientific interests. He worked in topology, differential geometry, calculus of variations, logic, graph theory and economics. Before the age of 30 he had written 65 papers. In 1937 he emigrated to the United States escaping from the deteriorated political and social atmosphere created in Austria by Hitler's regime. His name is attached, besides to a notion of curvature in general metric spaces, to the universal curve or Menger sponge [**BM**] and to some quantities, called Caley-Menger determinants [**B**], he used to characterize Euclidean spaces among metric spaces. The interested reader is referred to [**K**] where more detailed information on Menger's life and his mathematical work is provided and further references are given.

Given three distinct points in the plane, say z_1, z_2 and z_3 the Menger curvature associated to them is

$$c(z_1, z_2, z_3) = R^{-1},$$

where R is the radius of the circle passing through z_1, z_2 and z_3. If two of the three points coincide then we set $c(z_1, z_2, z_3) = 0$. In this way the condition $c(z_1, z_2, z_3) = 0$ is equivalent to the three points being collinear.

The picture below shows examples of triples having small curvature. In each case the reader is invited to imagine the circle passing through the three points and to realize that its radius is large. Indeed, the two pictures are essentially the same modulo a rotation and a dilation.

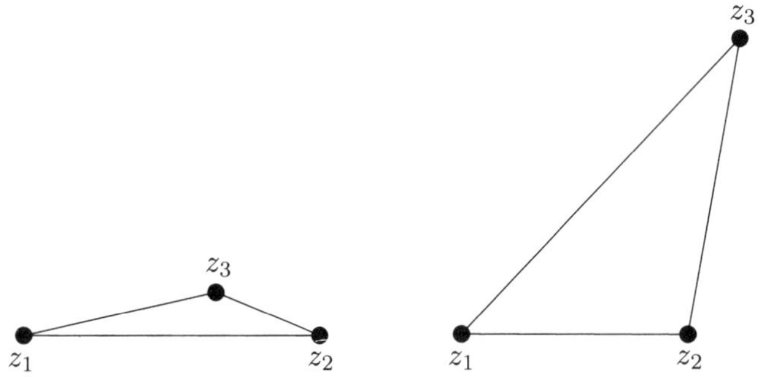

FIGURE 1

There are a couple of beautiful formulas for Menger curvature coming from elementary geometry that help in understanding the size of $c(z_1, z_2, z_3)$ in concrete examples. The first one is

$$c(z_1, z_2, z_3) = 2\,\frac{\sin\alpha}{l} \tag{2.1}$$

where l is the length of a side of the triangle T determined by z_1, z_2 and z_3 and α is the angle opposite to that side. The second formula, which can be deduced readily from (2.1), is

$$c(z_1, z_2, z_3) = \frac{4\,\mathrm{area}(T)}{|z_1 - z_2||z_1 - z_3||z_2 - z_3|}. \tag{2.2}$$

From (2.1) (or (2.2)) is completely obvious that the curvature of the two triples shown in Figure 1 becomes small provided z_1 and z_2 are kept fixed and z_3 tends, in the first picture, to a given point in between z_1 and z_2, and to infinity in the second picture. As a last example take $z_1 = -1$, $z_2 = 1$ and $z_3 = e^{i\epsilon}$, $\epsilon \in \mathbb{R}$. As ϵ tends to 0, $c(z_1, z_2, z_3)$ is always 1, in spite of the fact that z_3 gets very close to the straight line determined by z_1 and z_2.

From Heron's formula relating the area of a triangle to its perimeter, one can write down an expression for $c(z_1, z_2, z_3)$ involving only the mutual distances between z_1, z_2 and z_3. This remark is not going to be used in the sequel but it could be useful if one were interested in considering some of the problems we are dealing with in more general metric spaces. It also helps in getting an intuition of why Menger curvature and Caley-Menger determinants are so useful in characterizing Euclidean spaces among metric spaces.

The relationship between Menger curvature and the Cauchy kernel appears as follows. Take three points z_1, z_2 and z_3 in the complex plane and consider the Cauchy matrix associated to them

$$C = \begin{pmatrix} 0 & \frac{1}{z_1 - z_2} & \frac{1}{z_1 - z_3} \\ \frac{1}{z_2 - z_1} & 0 & \frac{1}{z_2 - z_3} \\ \frac{1}{z_3 - z_1} & \frac{1}{z_3 - z_2} & 0 \end{pmatrix}.$$

Let $\mathbf{1}$ denote the vector $(1, 1, 1) \in \mathbb{C}^3$, set

$$\langle z, w \rangle = \sum z_i \overline{w_i}, \text{ for } z, w \in \mathbb{C}^3, \tag{2.3}$$

and let C^* stand for the adjoint of C with respect to the standard hermitian product (2.3) in $\mathbb{C}^3$.

With the notation just established one gets

$$\|C(\mathbf{1})\|^2 = \langle C^* C(\mathbf{1}), \mathbf{1} \rangle = \sum_{i \neq j} |z_i - z_j|^{-2} + \sum_\sigma \frac{1}{(z_{\sigma(2)} - z_{\sigma(1)})\overline{(z_{\sigma(3)} - z_{\sigma(1)})}},$$

where in the second term the sum is over all permutations of $\{1, 2, 3\}$. Consequently, the square of the norm of the vector $C(\mathbf{1})$ is given by the sum of two terms, the first of which is the trace of C^*C and turns out to be a positive number, in accordance with the fact that C^*C is a positive matrix. The second term is the sum of the six entries of C^*C which are not on the main diagonal. Since C^*C is positive, this quantity must be real. The absolutely astonishing fact, for which the author has no explanation besides the blind computation performed below, is that the second term is a non-negative number. More precisely, we have the following.

LEMMA 2.1. *Given three points z_1, z_2 and z_3 in the plane one has*

$$\sum_\sigma \frac{1}{(z_{\sigma(2)} - z_{\sigma(1)})\overline{(z_{\sigma(3)} - z_{\sigma(1)})}} = c(z_1, z_2, z_3)^2, \tag{2.4}$$

where the sum is over all permutations of $\{1, 2, 3\}$.

The above identity was discovered by Melnikov in the course of his research on discrete expressions for analytic capacity [**Me**]. That (2.4) could be used in the *Calderón-Zygmund* theory of the Cauchy Integral was realized later by the author ([**V2**] and [**MV**]).

PROOF OF THE LEMMA. Coupling each term in (2.4) with its complex conjugate we see that the left hand side of (2.4) is equal to

$$2\sum_{i=1}^{3}\left(\frac{z_j - z_i}{|z_j - z_i|}, \frac{z_k - z_i}{|z_k - z_i|}\right)\frac{1}{|z_j - z_i||z_k - z_i|} \tag{2.5}$$

where (,) stands for the standard scalar product in $\mathbb{R}^2 = \mathbb{C}$, and, for each given i, j and k are the other two indices in $\{1, 2, 3\}$. Let l_i, $i = 1, 2, 3$, denote the lengths of the sides of the triangle determined by z_1, z_2 and z_3 and let θ_i be the angle opposite to the side of length l_i. Then (2.5) is equal to

$$2\left(\frac{\cos\theta_1}{l_2 l_3} + \frac{\cos\theta_2}{l_1 l_3} + \frac{\cos\theta_3}{l_1 l_2}\right),$$

which by (2.1) is

$$c(z_1, z_2, z_3)^2\,\frac{\sin(2\theta_1) + \sin(2\theta_2) + \sin(2\theta_3)}{4\sin\theta_1\sin\theta_2\sin\theta_3} = c(z_1, z_2, z_3)^2$$

where the last identity follows by eliminating θ_3 and working out elementary trigonometric formulas. □

3. The Cauchy Integral on Lipschitz graphs: background

In this section we introduce some background facts on the L^2 boundedness of the Cauchy Integral on a Lipschitz graph: we describe two possible routes to get to the problem and we briefly sketch some relevant parts of the long history of the proof.

Let Γ be a simple closed oriented rectifiable curve in the plane or a simple oriented locally rectifiable curve through infinity. One should have in mind the model examples of the unit circle $\{z : |z| = 1\}$ and the real axis. Because of rectifiability there is an arc-length measure on Γ, which we denote by $|dz| = ds$. Take $f \in L^1(ds)$ and consider its Cauchy Integral

$$Cf(z) = \frac{1}{2\pi i}\int_\Gamma \frac{f(\zeta)}{\zeta - z}\,d\zeta, \quad z \notin \Gamma. \tag{3.1}$$

We would like to understand the behaviour of $Cf(w)$ as the point $w \notin \Gamma$ tends to some point z on Γ. Thus our first task is to discuss the possibility of giving a sense to the integral in the right hand side of (3.1) for $z \in \Gamma$. It becomes clear immediately that the integral in (3.1) is not necessarily absolutely convergent if $z \in \Gamma$, because the kernel is not absolutely integrable with respect to ds. Because of this, one looks at the truncated integrals

$$C_\epsilon f(z) = \frac{1}{2\pi i}\int_{|\zeta - z| > \epsilon} \frac{f(\zeta)}{\zeta - z}\,d\zeta, \quad z \in \Gamma, \quad \epsilon > 0,$$

which are always absolutely convergent, and sets

$$Cf(z) = \lim_{\epsilon \to 0} C_\epsilon f(z), \quad z \in \Gamma, \tag{3.2}$$

whenever the limit exists. In other words, the integral in (3.1) should be understood in the principal value sense whenever $z \in \Gamma$ and so the problem of existence of principal values of Cauchy Integrals arises naturally. Now, if f has some smoothness, say f is the restriction to Γ of a function of class C^1 on the whole plane (compactly supported if Γ is unbounded), then (3.2) exists for all $z \in \Gamma$. To prove this simple fact, denote by Ω^+ and Ω^- the domains on the left and right hand sides of Γ respectively. The argument becomes easier if Γ is compact and so we concentrate on this case. Assume also that the orientation has been chosen so that Ω^+ is bounded. Setting $\gamma_\epsilon = \{\zeta : |\zeta - z| > \epsilon\} \cap \Gamma$ and $\sigma_\epsilon = \{\zeta : |\zeta - z| = \epsilon\} \cap \Omega^+$

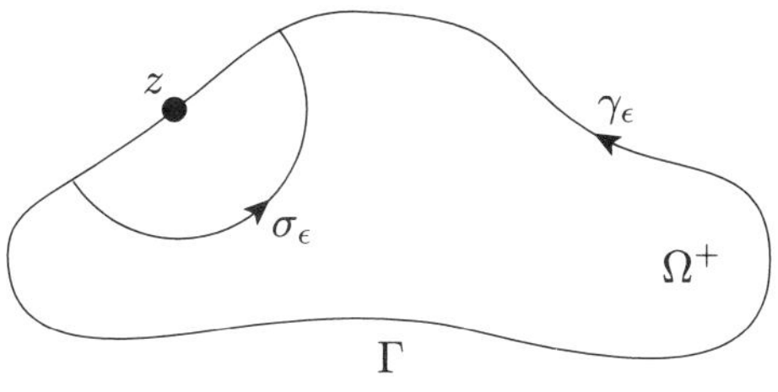

FIGURE 2

we have

$$\int_{\gamma_\epsilon} \frac{f(\zeta)}{\zeta - z}\, d\zeta = \int_{\gamma_\epsilon} \frac{f(\zeta) - f(z)}{\zeta - z}\, d\zeta + f(z) \int_{\sigma_\epsilon} \frac{d\zeta}{\zeta - z}$$

and thus, dividing by $2\pi i$ and letting $\epsilon \to 0$, we get

$$Cf(z) = \frac{1}{2\pi i} \int_\Gamma \frac{f(\zeta) - f(z)}{\zeta - z}\, d\zeta + \frac{f(z)}{2}, \tag{3.3}$$

provided Γ has a tangent at the point z.

Let us keep the smoothness assumption on f and consider what happens to $Cf(w)$ when $w \notin \Gamma$ tends to z. Notice that

$$Cf(w) = \frac{1}{2\pi i} \int_\Gamma \frac{f(\zeta) - f(w)}{\zeta - w}\, d\zeta + f(w), \quad w \in \Omega^+,$$

and

$$Cf(w) = \frac{1}{2\pi i} \int_\Gamma \frac{f(\zeta) - f(w)}{\zeta - w}\, d\zeta, \quad w \in \Omega^-.$$

If we let $C^\pm f(z)$ stand for the limits of $Cf(w)$ as w tends to z from $\Omega^\pm$, then by dominated convergence

$$C^+ f(z) = \frac{1}{2\pi i} \int_\Omega \frac{f(\zeta) - f(z)}{\zeta - z}\, d\zeta + f(z)$$

and

$$C^- f(z) = \frac{1}{2\pi i} \int_\Omega \frac{f(\zeta) - f(z)}{\zeta - z}\, d\zeta.$$

Comparing with (3.3) we obtain the elegant Plemelj's formulas (Sojotsky's in Russia)

$$\begin{cases} C^+ f(z) = Cf(z) + \frac{1}{2} f(z) \\ C^- f(z) = Cf(z) - \frac{1}{2} f(z) \end{cases}. \tag{3.4}$$

Indeed, we have proved (3.4) under the hypothesis that there is a tangent to Γ at the point z, but this occurs almost everywhere on Γ because of rectifiability.

What happens if f is a general function on $L^1(ds)$ without any additional smoothness properties? Experience with the model cases shows that then one has to restrict the way in which w approaches Γ. It turns out that only a non-tangential approach can be allowed if limits are required to exist. Fortunately, there is no problem at all in formalizing this notion for a general rectifiable curve, because tangents, and thus normals, exist at almost all points of the curve. Privalov was able to show that for $f \in L^1(ds)$ and for almost all $z \in \Gamma$ one has

$$Cf(w) - C_\epsilon f(z) \to \pm \frac{f(z)}{2}$$

as w tends non-tangentially to z from $\Omega^\pm$, where $\epsilon = 2|w - z|$.

Denote by $C^\pm f(z)$ the limits, if they exist, of $Cf(w)$ as w tends non-tangentially to z from $\Omega^\pm$. We conclude that a.e. on Γ the existence of one of the limits $Cf(z)$, $C^+ f(z)$, $C^- f(z)$ implies the existence of the other two and the validity of the formulas (3.4). The following basic question then arises:

(3.5) Does the principal value integral $Cf(z)$ exist almost everywhere for any f in $L^1(ds)$?

There is an argument to reduce the above problem to the case of C^1 curves that will be outlined briefly below. But assume for the moment that the reduction has been performed. Since a simple C^1 curve is locally a rotation of a C^1 graph we can suppose, without loss of generality, that Γ is the graph of a C^1 function A. Localizing and rotating again if necessary, we can further assume that the slopes of our graph are as small as we wish or, in other words, that $\|A'\|_\infty$ is arbitrarily small. After a short moment of reflection one concludes that there is no essential difference, for the kind of problem we are facing, between the assumptions of continuity and of boundedness of A'. As is well known, for a locally integrable function A the boundedness of A' and the Lipschitz condition

$$|A(x) - A(y)| \le M|x - y|, \quad x, y \in \mathbb{R},$$

are equivalent. Therefore our problem (3.5) has been reduced to Lipschitz graphs with an arbitrarily small Lipschitz constant M.

We now give an indication of how one can reduce the problem from the general rectifiable case to the C^1 case. A rectifiable curve has, by definition, a parametrization given by a function of bounded variation. By means of the arclength parameter s one finds a parametrization $z(s)$ with bounded $z'(s)$, which is not good enough because we are aiming at continuity of $z'(s)$, not just boundedness. Nevertheless, using Egoroff's Theorem, is not difficult to see that one can get the following Lusin type result: given $\epsilon > 0$ there is a closed subset E in the arclength parameter interval $[0, L]$ such that the Lebesgue measure of $[0, L]\backslash E$ is less than ϵ and the restriction of $z'(s)$ to E is continuous there. Moreover, one also gets

$$z(s) - z(t) = z'(t)(s - t) + o(s - t), \quad s, t \in E.$$

Whitney's extension theorem now provides a C^1 curve Γ' whose intersection with Γ is so big that $\Gamma\backslash\Gamma'$ has arclength measure less than ϵ. Thus Γ can be expressed as $\bigcup_{n=1}^{\infty} (\Gamma \cap \Gamma_n) \cup N$ where N has vanishing arclength and Γ_n is a C^1 curve. Using this one can complete the reduction without much difficulty.

We can now try to answer (3.5) when Γ is a Lipschitz graph (with small constant). To get more insight on the difficulties of the question we are considering, one can parametrize Γ by $\gamma(x) = x + iA(x)$, $x \in \mathbb{R}$, and write, for $z = \gamma(x)$,

$$\int_{|\zeta - z| > \epsilon} \frac{f(\zeta) - f(z)}{\zeta - z}\, d\zeta = \int_{|y-x|>\epsilon} \frac{f(\gamma(y))(1 + iA'(y))}{y - x + i(A(y) - A(x))}\, dy.$$

Absorbing the bounded factor $1 + iA'(y)$ in f, one is then led to inquire about the existence of the principal value integral

$$Tf(x) = \lim_{\epsilon \to 0} \int_{|y-x|>\epsilon} \frac{f(y)\, dy}{y - x + iA(y) - A(x)}, \tag{3.6}$$

for $f \in L^1(\mathbb{R})$. Since the kernel in (3.6) is odd, it is a simple matter to convince oneself, even without resorting to what we did before, that the limit in (3.6) exists for all x, provided f is a compactly supported function of class C^1 on $\mathbb{R}$. Now, there are well known classical real variable techniques that tell us that the a.e. existence of the principal value integral (3.6) follows from an estimate for the truncated operators

$$T_\epsilon f(x) = \int_{|y-x|>\epsilon} \frac{f(y)\, dy}{y - x + i(A(y) - A(x))}, \quad x \in \mathbb{R},$$

of the type

$$|\{x : |T_\epsilon f(x)| > t\}| \leq Ct^{-1}\|f\|_1, \quad t > 0, \tag{3.7}$$

with C independent of ϵ. In turn (3.7) follows, via the famous Calderón-Zygmund decomposition, from the L^2 estimate

$$\int_{-\infty}^{\infty} |T_\epsilon f(x)|^2\, dx \leq C \int |f(x)|^2\, dx, \tag{3.8}$$

with C again independent of ϵ. Hence Privalov's central question (3.5) can be answered in the positive sense, once one knows (3.8) to hold.

What we have done actually is reduce Privalov's problem to estimating in L^2 the singular integral operator of Calderón-Zygmund type T. Notice that T is not a convolution operator and so Fourier transform techniques seem to lose all their power.

On the other hand, since one is particularly interested in the case in which the graph of A is extremely close to the real axis, one can view the kernel of T as a very small perturbation of $\frac{1}{y-x}$, the kernel of the familiar Hilbert transform. This naive approach works when $A \in C^{1+\epsilon}$ but breaks down in the limiting case $\epsilon = 0$.

Another route that takes us directly to the problem of L^2 boundedness of the Cauchy Integral on Lipschitz graphs is the one discovered by Calderón. He was interested in constructing algebras of singular integral operators to develop a symbolic calculus that would then be applied to extend some fundamental results of classical PDE theory to equations with minimal smoothness requirements on their coefficients. This is very nicely explained in [**C3**] and [**S**].

If one looks at the composition of the simplest operators belonging to the class arising in Calderón's work in dimension one, then one is led to the commutator between the operator of multiplication by the Lipschitz function A and $H\frac{d}{dx}$, H being the Hilbert transform. It is a simple matter to write this operator as

$$C_1 f(x) = \left(AH\frac{d}{dx} - H\frac{d}{dx}A\right)(f) = \text{P.V.} \int_{-\infty}^{\infty} \frac{A(y) - A(x)}{(y-x)^2} f(y)\, dy, \tag{3.9}$$

where P.V. stands for principal value and f is a test function. The problem here is the $L^2(\mathbb{R})$ boundedness of C_1. This was achieved in 1965 by Calderón [**C1**] using complex analytic techniques involving the Hardy space H^1 and the Lusin area function. The higher order commutators

$$C_n f(x) = \text{P.V.} \int_{-\infty}^{\infty} \left(\frac{A(y) - A(x)}{y - x} \right)^n \frac{f(y)}{y - x} \, dy \tag{3.10}$$

were not amenable, however, to these methods. Notice that the Cauchy Integral (3.6) and the family of commutators (3.10) can be written as

$$Tf(x) = \text{P.V.} \int_{-\infty}^{\infty} F\left(\frac{A(y) - A(x)}{y - x} \right) \frac{f(y)}{y - x} \, dy$$

for appropriate choices of F, namely $F(t) = (1 + it)^{-1}$ or $F(t) = t^n$, and thus, in some sense, they are close relatives. In fact, if $\|A'\|_\infty < 1$ then, expanding the kernel of the Cauchy Integral, we have

$$Tf(x) = \sum_{n=0}^{\infty} (-i)^n C_n f(x). \tag{3.11}$$

If you are able to prove that C_n is bounded on $L^2(\mathbb{R})$ with an estimate on its norm of the type $\|C_n\| \leq C^n \|A'\|_\infty^n$, then you can sum up the series (3.11) and show the boundedness of T, at least for graphs satisfying $C\|A'\|_\infty < 1$.

To get boundedness of the Cauchy Integral without any restriction on $\|A'\|_\infty$ you would need to show polynomial estimates of the kind

$$\|C_n\| \leq C_0 (1 + n)^k \|A'\|_\infty^n,$$

where k is a positive integer independent of n. This is exactly what Coifman, McIntosh and Meyer did in 1982 (with $k = 9$) in their famous paper [**CMM**]. Five years before, in 1977, Calderón had been able to prove, again by complex analytic methods, that the Cauchy Integral is bounded for Lipschitz graphs with sufficiently small constant, thus solving Privalov's problem [**C2**] and [**C3**]. Around that time (and later) remarkable achievements were obtained by Coifman and Meyer on commutators related to pseudodifferential operators, paraproducts and multilinear singular integrals.

4. L^2 boundedness of the Cauchy Integral on a Lipschitz graph

In this section we discuss the author's proof of the L^2 boundedness of the Cauchy Integral T on the graph of the Lipschitz function $y = A(x)$ (see [**V2**] and [**MV**]).

Given an interval I and a function $f \in L^2(I)$ one starts by trying to find a good expression for the norm of $T_\epsilon f$ in $L^2(I)$. It is possible to obtain the formula (χ_I is the characteristic function of I)

$$2 \int_I |T_\epsilon(f)|^2 + 4 \operatorname{Re} \int_I T_\epsilon(f) \overline{T_\epsilon(\chi_I)} f = G + M, \tag{4.1}$$

where G is a term of a geometric nature, containing basically Menger curvature, and M is a very good term under control. In fact M is closely related to the Hardy-Littlewood maximal operator (see [**V3**]). It is also important to notice that the second term on the left hand side of (4.1) is harmless, because of the Cauchy-Schwarz inequality, once one knows how to estimate the norm of $T_\epsilon(\chi_I)$ in $L^2(I)$; for more details on this see the argument that leads to (4.7) below. To feel the potential force of (4.1) let us look at the case $A \equiv 0$, so that the graph is just the real axis: then one can see that G vanishes, because a straight line has zero curvature, and that (4.1) yields

$$\int_I (H_\epsilon f)^2 \leq C \int_I (Mf)^2,$$

where $H_\epsilon = T_\epsilon$ is the truncated Hilbert transform and Mf the Hardy-Littlewood maximal operator applied to f. The idea is then that in the general case the term G will have the appropriate control thanks to Menger curvature, which one hopes to be related to the geometry of the graph.

In the particularly important special case $f = \chi_I$ (4.1) reduces to

$$6 \int_I |T_\epsilon(\chi_I)|^2 = G + M \tag{4.2}$$

and M can be shown to be to less than $C|I|$.

Let us proceed to discuss in some detail the case $f = \chi_I$. Parametrizing the graph by $\gamma(x) = x + iA(x)$ we have

$$\begin{aligned}\int_I |T_\epsilon(\chi_I)|^2 &= \int_I T_\epsilon(\chi_I)(x)\overline{T_\epsilon(\chi_I)(x)}\,dx \\ &= \int_I \int_{I_\epsilon(x)} \int_{I_\epsilon(x)} \frac{dx\,dy\,dz}{(\gamma(y)-\gamma(x))\overline{(\gamma(z)-\gamma(x))}},\end{aligned}$$

where $I_\epsilon(x) = \{t \in I : |t-x| > \epsilon\}$. In order to symmetrize the domain we set

$$S_\epsilon = \{(x,y,z) \in I^3 : |y-x| > \epsilon,\ |z-x| > \epsilon,\ |z-y| > \epsilon\}.$$

Standard estimates show that

$$\int_I |T_\epsilon(\chi_I)|^2 = \iiint_{S_\epsilon} \frac{dx\,dy\,dz}{(\gamma(y)-\gamma(x))\overline{(\gamma(z)-\gamma(x))}} + E$$

where E is an error term satisfying $|E| \leq C|I|$. By permutating the positions of the three variables in the integrand of the right hand side of the preceding formula we get 6 different expressions for the left hand side. Using Lemma 2.1 we then see that

$$\int_I |T_\epsilon(\chi_I)|^2 = \frac{1}{6} \iiint_{S_\epsilon} c^2(\gamma(x),\gamma(y),\gamma(z))\,dx\,dy\,dz + O(|I|), \tag{4.3}$$

which is (4.2) if we let G be the triple integral above.

Hence it is enough to prove the inequality

$$\iiint_{I^3} c^2(\gamma(x),\gamma(y),\gamma(z))\,dx\,dy\,dz \leq C\|A'\|_\infty^2 |I|. \tag{4.4}$$

The area of the triangle with vertices $\gamma(x)$, $\gamma(y)$ and $\gamma(z)$ is

$$|(A(y)-A(x))(z-x) - (A(z)-A(x))(y-x)|$$

and thus, because of (2.2),

$$c(\gamma(x),\gamma(y),\gamma(z)) \leq 4 \left| \frac{\dfrac{A(y)-A(x)}{y-x} - \dfrac{A(z)-A(x)}{z-x}}{z-y} \right|. \tag{4.5}$$

We need now the following lemma whose proof is not much more than a beautiful exercise with Plancherel's Theorem.

LEMMA 4.1. *If a is a locally integrable function with derivative $a' \in L^2(\mathbb{R})$, then*

$$\int_{-\infty}^{\infty}\int_{-\infty}^{\infty}\int_{-\infty}^{\infty} \left| \frac{\frac{a(y)-a(x)}{y-x} - \frac{a(z)-a(y)}{z-y}}{z-x} \right|^2 dx\,dy\,dz = c_0 \int_{-\infty}^{\infty} |a'(x)|^2\,dx,$$

c_0 being a numerical constant.

Our next step will be to localize the L^2 identity provided by the lemma to the given interval I. This is a kind of technique one uses often when working with BMO. Indeed, in the next inequality the BMO norm of A' will show up naturally.

Apply the lemma to the function $a = \chi_I(A - P_I)$, where $P_I(x) = A'_I(x-\alpha) + A(\alpha)$, $A'_I = \frac{1}{|I|}\int_I A'$, $I = (\alpha, \beta)$. The result is

$$\iiint_{I^3} c^2(\gamma(x),\gamma(y),\gamma(z))\,dx\,dy\,dz \le C\int_I |A' - A'_I|^2 \le C\|A'\|_\infty^2 |I|,$$

which completes the proof of the estimate

$$\int_I |T_\epsilon(\chi_I)|^2 \le C(1 + \|A'\|_\infty^2)|I|, \text{ for all intervals } I. \tag{4.6}$$

If one is willing to apply the $T(1)$-Theorem of David and Journé, the proof is complete. Otherwise, there is a simple way of avoiding $T(1)$ which we now describe. Take a (real) function $b \in L^\infty(I)$ and try to express the $L^2(I)$ norm of $T_\epsilon(b)$ in terms of $c^2(\gamma(x),\gamma(y),\gamma(z))$ using the permutation trick described above. One can do that easily and one obtains the identity

$$\begin{aligned} 2\int_I |T_\epsilon(b)|^2 &+ 4\,\mathrm{Re}\int_I T_\epsilon(b)\overline{T_\epsilon(\chi_I)}b \\ &= \iiint_{S_\epsilon} c^2(\gamma(x),\gamma(y),\gamma(z))b(x)b(y)\,dx\,dy\,dz + O\left(\int_I |b|^2\right), \end{aligned}$$

which is a concrete version of (4.1) with f replaced by b.

Hence, using (4.4) and (4.6),

$$\begin{aligned} \int_I |T_\epsilon(b)|^2 \le C(1 + \|A'\|_\infty^2)^{1/2}\left(\int_I |T_\epsilon(b)|^2\right)^{1/2} \|b\|_\infty |I|^{1/2} \\ + C(1 + \|A'\|_\infty^2)\|b\|_\infty^2 |I|, \end{aligned}$$

which clearly gives

$$\int_I |T_\epsilon(b)|^2 \le C(1 + \|A'\|_\infty^2)\|b\|_\infty^2 |I|. \tag{4.7}$$

Now (4.7) implies that T_ϵ sends boundedly $H^1(\mathbb{R})$ into $L^1(\mathbb{R})$ (use atoms) and $L^\infty(\mathbb{R})$ into $\mathrm{BMO}(\mathbb{R})$. By interpolation we finally obtain that T_ϵ maps boundedly $L^2(\mathbb{R})$ into $L^2(\mathbb{R})$ (in fact $L^p(\mathbb{R})$ into $L^p(\mathbb{R})$, $1 < p < \infty$).

In the next section we will show that the quantity in the right hand side of (4.5) is intimately connected to the first Calderón commutator C_1. In fact it turns out that (4.3) and (4.5) say that the Cauchy Integral is dominated by the first commutator. Thus in some ironic sense, the 1965 paper **[C1]** is the beginning and almost the end of the story. Of course, this was far from being even conceivable in the sixties and, indeed, much clever and hard work had to be done for many years to fully understand the Cauchy Integral. The beautiful quotation from Montaigne that opens the book by Meyer and Coifman [**MC**] suits perfectly, I believe, to the long and complicated process of understanding the Cauchy Integral.

5. L^2 boundedness of the first Calderón commutator

In this section we present a simple proof of the L^2 boundedness of the first Calderón commutator (3.9) and we show that the Cauchy Integral is controlled by the first commutator. As the reader will see, we basically reinterpret the ideas and results introduced in the preceding section. Other proofs of the L^2 boundedness of the first commutator can be found in [**Mu**].

Call

$$K(x,y)=\frac{A(y)-A(x)}{(y-x)^2}$$

the kernel of the first commutator. We want to exploit further the symmetrization idea described before.

Write

$$C_{1,\epsilon}(f)(x)=\int_{|y-x|>\epsilon}K(x,y)f(y)\,dy, \quad x\in\mathbb{R}.$$

Given an interval I, we have

$$\begin{aligned}\int_I C_{1,\epsilon}(\chi_I)^2&=\int_I\int_{I_\epsilon(x)}\int_{I_\epsilon(x)}K(x,y)K(x,z)\,dx\,dy\,dz\\&=\iiint_{S_\epsilon}K(x,y)K(x,z)\,dx\,dy\,dz+O(|I|),\end{aligned}$$

where $I_\epsilon(x)$ and S_ϵ are defined in the preceding section. Notice that the integrand in the triple integral above is already symmetric in y and z. If we interchange x by y, and x by z, we get two new different expressions for $\int_I C_{1,\epsilon}(\chi_I)^2$. Therefore

$$\int_I C_{1,\epsilon}(\chi_I)^2=\frac{1}{3}\iiint_{S_\epsilon}S(x,y,z)\,dx\,dy\,dz+O(|I|)$$

where

$$S(x,y,z)=K(x,y)K(x,z)+K(y,x)K(y,z)+K(z,y)K(z,x).$$

It turns out that, as a straightforward computation shows,

$$S(x,y,z)=\left(\frac{\dfrac{A(y)-A(x)}{y-x}-\dfrac{A(z)-A(y)}{z-y}}{z-x}\right)^2,$$

and so Lemma 4.1 gives, after localization,

$$\int_I C_{1,\epsilon}(\chi_I)^2\le C(1+\|A'\|_\infty^2)|I|.$$

The L^2 boundedness is again a consequence of the $T(1)$-Theorem or of the (simpler) fact that one can interpolate between H^1 and BMO. On the other hand (4.3) and (4.5) yield

$$\int_I |T_\epsilon(\chi_I)|^2\le C\int_I C_{1,\epsilon}(\chi_I)^2+O(|I|),$$

thus showing that the Cauchy Integral can be controlled by the first commutator.

Joan Mateu and the author have recently applied the symmetrization technique to the n-th Calderón commutator obtaining what seems to be the best possible asymptotic estimate

$$\|C_n\| \leq C_0(1+n)\|A'\|_\infty^n.$$

6. The solution of Vitushkin's conjecture

Menger curvature has been one of the key ingredients in the proof of Vitushkin's conjecture, which has been recently completed by G. David [**D2**]. When one considers the proof in its totality one realizes that it is an impressive achievement, because it involves several deep ideas of different nature that various people have been introducing at successive stages, sometimes for seemingly unrelated purposes. Some parts of the proof are still too involved technically and some further work seems necessary to perceive more clearly the real importance of each of the ideas involved and purify the technical development of the proof. A good step in this direction has been taken by Nazarov, Treil and Volberg in the nice article [**NTV2**]. An excellent exposition by G. David, including a historical perspective, is available at present [**D3**]. After reading that paper you get the feeling that no human being could have proved the result even two decades after it was conjectured because too many essential steps were still missing. Although this is a well known and common phenomenon in mathematics, is still impressive to perceive it directly in a concrete instance that you can seize, at least approximately.

In this section we will try to give an idea, even if vague, of some of the main steps in the proof of the conjecture.

Vitushkin's conjecture is the solution, in a particular case, of Painlevé's problem. Painlevé proved that planar compact sets of zero length (one dimensional Hausdorff measure) are removable for bounded analytic functions. Removability of a compact set E means that if an open set Ω contains E and $f \in H^\infty(\Omega \backslash E)$, then f can be continued analytically to Ω. In fact, it is not difficult to see that one only needs to check the above condition for $\Omega = \mathbb{C}$ and so (by Liouville) that E is removable if and only if $H^\infty(\mathbb{C}\backslash E)$ is reduced to constants. The first trace of the notion of removability goes back to Riemann who proved, as we all know, that a point is removable. Painlevé asked, more than one hundred years ago, if removable sets can be described in geometric or metric terms, and nobody has been able yet to provide an answer to a question of such an innocous appearance.

That Painlevé's result is, in some sense, sharp becomes apparent by considering an interval I on the real axis: I has positive length and is not removable, because the conformal mapping from the complement of I in the extended complex plane onto the unit disc cannot be continued analytically through I.

Vitushkin discovered that the converse of Painlevé's result is not true, that is, that you can build a removable compact set of positive length. The simplest possible example of such a phenomenon was found independently by Garnett and Ivanov. It turns out to be the planar Cantor set one generates by the procedure of taking "corner quarters". Since it is an important object for us, we briefly review its definition.

We start with the unit square $Q = [0, 1] \times [0, 1]$. We divide Q into sixteen squares of side length $1/4$ and we take those four that contain the vertices of Q. We repeat the procedure in each of these four squares and we get sixteen squares of side length 1/16. Proceeding inductively we have at the n-th generation 4^n squares of side length 4^{-n}. Let E_n denote their union and set $E = \bigcap_{n=1}^{\infty} E_n$, the Cantor set we are interested in.

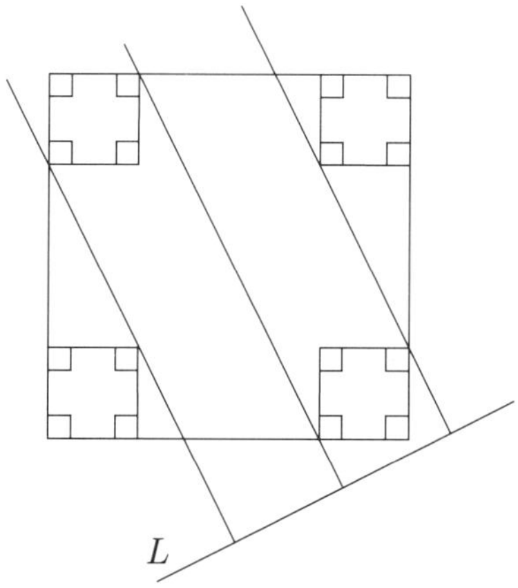

FIGURE 3

The orthogonal projection of each E_n, and consequently of E, on the line L shown in Figure 3 is some interval (the same for each n). Therefore the length of E is positive because a projection does not increase length. Moreover, it is easily seen that E has finite length. It can also be shown, and this is not obvious, that E is removable [**G1**].

The right question is now the following: what smallness property does the set E enjoy that implies removability and that is not shared by an interval on the real line? An indication of what the answer might be is given by Besicovitch theory of (measurable) sets of finite length [**F**] and [**Ma**]. Besicovitch showed in the thirties that any set of finite length E can be expressed as $E = N \cup R \cup I$ where N has vanishing length, and R and I are (either the empty set or) sets of positive length of opposite extremal nature. The set R is rectifiable, that is, is a subset of a rectifiable curve. The set I is irregular, in the sense that intersects all rectifiable curves in zero length. Alternatively, in spite of having positive length, I projects into sets of zero length in almost all directions. It turns out that the Cantor set we considered before is irregular and this fact explains why its behaviour, as far as removability is concerned, is so different from that of an interval.

Vitushkin's conjecture now becomes natural. It reads as follows [**Vi**].

If E is a compact set of positive finite length, then E is removable if and only if E is irregular in Besicovitch's sense.

Notice that Besicovitch irregularity is invariant by bilipschitz mappings and thus is a metric condition.

Calderón's contribution is the proof of the necessity of Besicovitch irregularity. Indeed, assume that E is not irregular. Then E intersects some rectifiable curve Γ in positive length. In fact Γ can be taken to be a C^1 graph with small constant because of the reduction argument described in section 3. Set $K = E \cap \Gamma$

and consider $f(z) = C(\chi_K)(z)$, $z \notin K$. Unfortunately f is not bounded, but, appealing to the L^2 boundedness of the Cauchy Integral on Γ, it can be seen to be in $H^2(\mathbb{C}\backslash K)$. An analytic function on $\mathbb{C}\backslash K$ is said to be in $H^2(\mathbb{C}\backslash K)$ if its L^2 norms on a certain sequence of curves surrounding and approximating K are uniformly bounded. Thus f is in $H^2(\mathbb{C}\backslash K)$ and cannot be continued analytically through K, because otherwise it would vanish identically by Liouville. That f is not identically zero becomes clear from the fact that $\lim_{z\to\infty} zf(z) = \text{length}(K) \neq 0$.

This is enough to show the non-removability of K, and thus that of E, thanks to a deep theorem of Garabedian (see [**G2**]) stating that the removable sets for H^∞ and H^2 are the same. It is worth mentioning that this is basically the argument to prove the Denjoy conjecture from Calderón's Theorem on the Cauchy Integral (see [**M**] and [**V1**] for more details).

Hence proving the sufficiency in the statement of Vitushkin's conjecture is the real difficult problem. One has to show that if E is non-removable, then there exists a rectifiable curve that intersects E in positive length. If one thinks for a while at how such a curve could be constructed, one realizes that in fact it is not obvious at all even how to begin.

The beta numbers of Peter Jones provide a hint of what could be tried. Introduced to deal with the Cauchy Integral on a Lipschitz graph [**J1**], the beta numbers play an essential role in deciding whether or not a given set is contained in a rectifiable curve, a question related to the famous traveling salesman problem of computational geometry [**J2**].

For a compact set E and a dyadic square Q, the beta number associated to E and Q is

$$\beta(Q) = \inf_L \sup_{z\in E\cap Q} \operatorname{dist}(z, L)\ell(Q)^{-1}, \tag{6.1}$$

where the infimum is taken over all straight lines L intersecting Q.

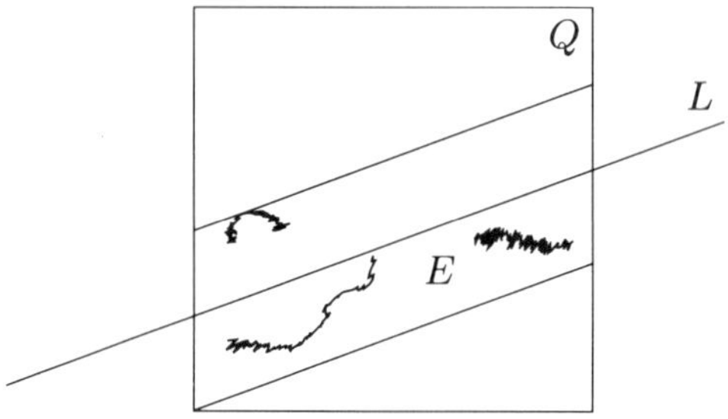

FIGURE 4

Therefore $\beta(Q)$ measures, in a scale invariant way, how much the set E deviates from a line at the scale and location determined by Q. It turns out that there exists a rectifiable curve Γ containing E if and only if $\sum_Q \beta^2(Q)\ell(Q)$ is finite, where the sum is taken over all dyadic squares [**J2**].

We are now much better equipped to confront Vitushkin's conjecture. Indeed, given a non-removable set E we have at our disposal the beta numbers to identify a rectifiable piece inside E. It is still unclear how one can get any information from non-removability. **M. Christ** made an important contribution in this direction [**Ch**]. In his original result he had to impose an additional requirement on E, namely that E has positive finite length in a uniform fashion:

$$C^{-1}r \leq \operatorname{length}(E \cap D(z,r)) \leq Cr, \quad z \in E, \quad 0 < r \leq \operatorname{diam}(E),$$

where $D(z,r)$ is the open disc of radius r centered at z. The above condition, called Ahlfors regularity, had been shown by G. David to characterize those rectifiable curves in the plane for which the Cauchy Integral is bounded on $L^2(ds)$ [**D1**]. Before stating Christ's result we need to establish some terminology. Take a positive Radon measure μ in the plane, and a positive number ϵ. For as compactly supported function f on $L^2(\mu)$ define

$$C_\epsilon(f\mu)(z) = \int_{|\zeta - z|>\epsilon} \frac{f(\zeta)}{\zeta - z}\, d\mu(\zeta), \quad z \in \mathbb{C}.$$

Then one says that the Cauchy Integral is bounded on $L^2(\mu)$ provided

$$\int |C_\epsilon(f\mu)|^2 \, d\mu \leq C \int |f|^2 \, d\mu, \tag{6.2}$$

with C independent of ϵ and f. If μ has no atoms, then a necessary condition for (6.2) is

$$\mu(D(z,r)) \leq Cr, \quad z \in \mathbb{C}, \quad r > 0. \tag{6.3}$$

For example, if μ is the length measure restricted to the Cantor set of Figure 3 then (6.3) holds but it can be proved that (6.2) does not.

THEOREM 6.1 (Christ). *If E is a non-removable Ahlfors regular compact set then there exists another Ahlfors regular compact set E' such that* $\operatorname{length}(E \cap E') > 0$ *and the Cauchy Integral is bounded on $L^2(\mu)$, where μ is length (one dimensional Hausdorff measure) restricted to $E \cap E'$.*

Hence non-removability is equivalent, at least for the class of Ahlfors regular sets, to L^2 boundedness of the Cauchy Integral with respect to the length measure on a non-trivial piece of the set (non-removability follows from L^2 boundedness by a variation of the argument we used in the proof of the necessary condition in Vitushkin's conjecture).

The reason for requiring Ahlfors regularity is that the proof uses an extremely strong criterion for L^2 boundedness of singular integrals of Calderón-Zygmund type known as the $T(b)$-Theorem, which is due to David, Journé and Semmes. In the original formulation the $T(b)$-Theorem required the underlying space to be of homogeneous type, that is, a metric space endowed with a Borel positive measure μ satisfying the doubling condition

$$\mu(2B) \leq C\mu(B), \text{ for all balls } B. \tag{6.4}$$

The length measure on an Ahlfors regular set is clearly doubling, but one finds without difficulty examples of sets of finite length for which the length measure restricted to the set does not satisfy (6.4).

If the kernel of the operator T is odd the $T(b)$-Theorem basically asserts that T is bounded on L^2 provided $T(b)$ is in BMO (in particular in L^∞) for some bounded function b satisfying a non-triviality condition called para-accretivity:

$$\frac{1}{\mu(B)} \left| \int_B b \, d\mu \right| \geq \delta > 0, \text{ for all balls } B. \tag{6.5}$$

We are now ready to get an idea of what non-removability has to do with the $T(b)$-Theorem. Assume that E is non-removable (and Ahlfors regular). Then there is a non-constant bounded holomorphic function B on $\mathbb{C}\setminus E$, which can be seen to be of the form

$$B(z) = \int_E \frac{b(\zeta)}{\zeta - z}\, d\mu(\zeta), \quad z \notin E,$$

μ being the length measure on E and b some function in $L^\infty(\mu)$.

When z tends to some point in E we are in trouble, because in this general context there is nothing like the Plemelj's formula. However, one still has

$$|C_\epsilon(b\mu)(z)| \leq C\|B\|_\infty, \quad z \in E, \quad \epsilon > 0,$$

and this can be understood, for each ϵ and in a uniform way, as "$T(b)$ is bounded" for the operator defined by $T(f) = C_\epsilon(f\mu)$.

Although the non-constancy of B translates into $\int_E b\, d\mu \neq 0$, b does not need to have locally a non zero integral in a uniform way, as in (6.5). In other words, b is not necessarily para-accretive and this justifies the need for a modification process, based on a stopping time argument, that in the end will transform E into E'. How Ahlfors regularity can be dispensed with will be mentioned later on.

Therefore, to complete the proof of Vitushkin's conjecture we need to establish a link between L^2 boundedness of the Cauchy Integral and Jones' beta numbers. It turns out that Menger curvature is the perfect technical device to do this. In the Ahlfors regular case the argument is rather transparent [**MMV**] if one is willing to use the David-Semmes theory of uniform rectifiability [**DS**]. Although there is now a direct way of relating Menger curvature to the beta numbers, we will sketch the original one. See [**Fa**] for the best result known before the Menger era.

Let E be an Ahlfors regular set. A first fact, of a technical nature but extremely useful, is that one has grids of "dyadic squares" in E, as in the Euclidean setting. A dyadic square in our context is a subset of E with no particular geometric shape. To understand what we mean it is convenient to consider the example of the Cantor set described at the beginning of this section. The dyadic squares of size 4^{-n} are the intersections of E with each of the 4^n squares appearing at the n-th generation. Then they have diameter and length comparable to 4^{-n}.

The dyadic squares of a fixed size form a partition of E and one still has the basic fact that if two given dyadic squares are not disjoint then one is contained in the other. The reader is referred to [**DS**] for complete details.

Let μ be the length measure on E and assume that the Cauchy Integral is bounded on $L^2(\mu)$, that is, that (6.2) holds. **David and Semmes** proved that E is (locally) contained in a rectifiable curve provided an additional condition, called the **weak geometric lemma**, holds [**DS**]. The result is not at all easy to prove, on the contrary, it is the product of a rather sophisticated strategy involving what has been called "Corona decompositions". To formulate the weak geometric lemma, set, for any given dyadic square Q_0 in E and any $\epsilon > 0$,

$$D(Q_0, \epsilon) = \{Q : Q \text{ dyadic square} \subset Q_0 \text{ such that } \beta(Q) > \epsilon\},$$

where $\beta(Q)$ is defined (essentially) as in (6.1) replacing $\ell(Q)$ by $\operatorname{diam}(Q)$. The weak geometric lemma is the following assertion, which can be true or not depending on the nature of the given Ahlfors regular set E.

For each dyadic square Q_0 and each $\epsilon > 0$ one has

$$\sum_{Q \in D(Q_0,\epsilon)} \mu(Q) \leq C\mu(Q_0). \tag{6.6}$$

The Carleson type condition (6.6) is clearly a weaker form of the "geometric lemma"

$$\sum_{Q \subset Q_0} \beta^2(Q)\,\mathrm{diam}(Q) \leq C\,\mathrm{diam}(Q_0),$$

that looks like a variation of the Jones condition.

It turns out that with the help of Menger curvature one can prove [**MMV**] that the weak geometric lemma is a consequence of the boundedness of the Cauchy Integral on $L^2(\mu)$. To explain this we start by noticing that playing the permutations game that we described in section 4, one can get without any additional effort

$$\int_D |C_\epsilon(\chi_D\mu)|^2\,d\mu = \iiint_{S_\epsilon} c^2(z,w,\zeta)\,d\mu(z)\,d\mu(w)\,d\mu(\zeta) + O(\mu(D)),$$

where D is a disc and

$$S_\epsilon = \{(z,w,\zeta) \in D^3 : |z-w| > \epsilon,\ |z-\zeta| > \epsilon \text{ and } |w-\zeta| > \epsilon\}.$$

If the Cauchy Integral is bounded on $L^2(\mu)$, then

$$\iiint_{D^3} c^2(z,w,\zeta)\,d\mu(z)\,d\mu(w)\,d\mu(\zeta) \leq C\mu(D), \text{ for each disc } D. \tag{6.7}$$

To check (6.6), let Q be a dyadic square in E, and take two points $a_1, a_2 \in Q$ such that

$$d \equiv |a_1 - a_2| \simeq \mathrm{diam}(Q) \simeq \mu(Q).$$

Let L be the straight line passing through a_1 and a_2, and let a_3 be a point in Q such that $h \equiv \mathrm{dist}(a_3, L) = \sup_{z\in Q} \mathrm{dist}(z, L)$. Recall that $Q \subset E$ and so h may be much smaller than d. Set $D_i = D(a_i, \frac{h}{3})$, $1 \leq i \leq 3$. Then we have

$$c(z_1,z_2,z_3) \geq Chd^{-2},$$

for some positive constant C and all $(z_1,z_2,z_3) \in D_1 \times D_2 \times D_3 \equiv T(Q)$.

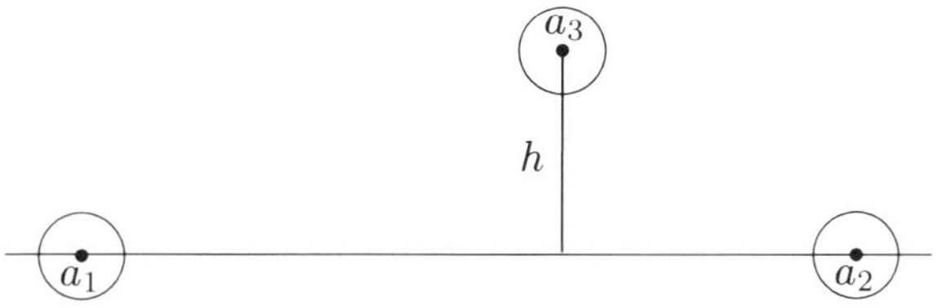

FIGURE 5

Assume now that $Q \subset Q_0$ and $\beta(Q) > \epsilon$. Since clearly $Chd^{-1} \geq \beta(Q) > \epsilon$,

$$\begin{aligned}\iiint_{T(Q)} c^2(z_1, z_2, z_3)\, d\mu(z_1)\, d\mu(z_2)\, d\mu(z_3) &\geq C(hd^{-2})^2 h^3 \\ &= Ch^5 d^{-4} \\ &\geq C\epsilon^5 d \\ &\geq C\epsilon^5 \mu(Q).\end{aligned}$$

It is not difficult to convince oneself that the $T(Q)$ have finite overlapping as Q varies.

Hence

$$\begin{aligned}\sum_{Q \in D(Q_0, \epsilon)} \mu(Q) &\leq C\epsilon^{-5} \sum_{Q \subset Q_0} \iiint_{T(Q)} c^2(z_1, z_2, z_3)\, d\mu(z_1)\, d\mu(z_2)\, d\mu(z_3) \\ &\leq C\epsilon^{-5} \iiint_{(4Q_0)^3} c^2(z_1, z_2, z_3)\, d\mu(z_1)\, d\mu(z_2)\, d\mu(z_3) \\ &\leq C\epsilon^{-5} \mu(Q_0),\end{aligned}$$

where the last inequality follows from (6.7). This completes the proof of (6.6).

The main steps in the proof of Vitushkin's conjecture in the Ahlfors regular case can be summarized in the following scheme.

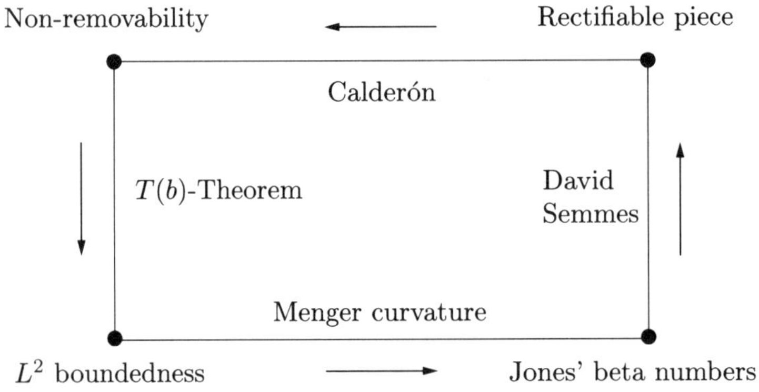

FIGURE 6

Getting rid of the Ahlfors regularity assumption requires new ideas in order to overcome the considerable difficulties caused by the lack of homogeneity. In particular one needs a $T(b)$-Theorem without the assumption that the underlying measure satisfies the doubling condition. That this could be proved looked rather unlikely at first glance, because one was used to believe that spaces of homogeneous type were the right environment in which Calderón-Zygmund Theory could be developed. Several people (David, Nazarov, Mattila, Tolsa, Treil and Volberg among others) have contributed during the last few years to show that, surprisingly enough, substantial parts of classical Calderón-Zygmund Theory hold without the doubling condition (see, for example, [**T1**], [**T2**], [**NTV1**] and [**D3**]). In particular the $T(b)$-Theorem holds in the non-homogeneous setting, as David proved in [**D2**]. For more details on the difficulties one has to face in the non-doubling context the reader is invited to consult the nice survey paper [**D3**] and the preprints [**NTV2**], [**NTV3**].

ACKNOWLEDGEMENT. The author was partially supported by the grants BFM2000-0361 and 2000SGR 00059.

References

[B] L. M. Blumenthal, *"Theory and applications of distance geometry"*, Chelsea, New York, 1970.

[BM] L. M. Blumenthal and K. Menger, *"Studies in Geometry"*, W. H. Freeman and Company, San Francisco, 1970.

[C1] A. P. Calderón, Commutators of singular integral operators, *Proc. Nat. Acad. Sci. U.S.A.* **53** (1965), 1092–1099.

[C2] A. P. Calderón, Cauchy Integrals on Lipschitz curves and related operators, *Proc. Nat. Acad. Sci. U.S.A.* **74** (1977), 1324–1327.

[C3] A. P. Calderón, Commutators, singular integrals on Lipschitz curves and applications, Proceedings of the International Congress of Mathematicians, Helsinki, 1978, pp. 85–96.

[Ch] M. Christ, A $T(b)$ theorem with remarks on analytic capacity and the Cauchy Integral, *Colloq. Math.* **60/61** (1990), 1367–1381.

[CM] R. R. Coifman and Y. Meyer, Au delà des opérateurs pseudo-differentials, *Astérisque* **57**, Société Mathématique de France (1978).

[CMM] R. R. Coifman, A. McIntosh and Y. Meyer, L'intégrale de Cauchy définit un opérateur borné sur L^2 pour les courbes lipschitziennes, *Ann. of Math.* **116** (1982), 361–387.

[D1] G. David, Opérateurs intégraux singuliers sur certaines courbes du plan complexe, *Ann. Sci. École Norm. Sup.* **17** (1984), 157–189.

[D2] G. David, Unrectifiable 1-sets have vanishing analytic capacity, *Rev. Mat. Iberoamericana* **14(2)** (1998), 369–479.

[D3] G. David, Analytic capacity, Calderón-Zygmund operators, and rectifiability, *Publ. Mat.* **43** (1999), 3–25.

[DS] G. David and S. Semmes, *"Analysis of and on uniformly rectifiable sets"*, AMS series of Mathematical surveys and monographs **38**, 1993.

[F] K. Falconer, *"The geometry of fractal sets"*, Cambridge Univ. Press, 1984.

[Fa] X. Fang, The Cauchy Integral, analytic capacity and subsets of quasicircles, *Pacific J. Math.* **166** (1994), 247–294.

[G1] J. Garnett, Positive length but zero analytic capacity, *Proc. Amer. Math. Soc.* **21** (1970), 696–699.

[G2] J. Garnett, *"Analytic capacity and measure"*, Springer-Verlag, 1972.

[J1] P. Jones, Square functions, Cauchy integrals, analytic capacity, and harmonic measure, in: *"Proc. Conf. on Harmonic Analysis and Partial Differential Equations"*, El Escorial 1987 (J. García-Cuerva, ed.), Lecture Notes in Math. **1384**, Springer-Verlag, 1989, pp. 24–68.

[J2] P. Jones, Rectifiable sets and the traveling salesman problem, *Invent. Math.* **102** (1990), 1–16.

[K] S. Kass, Karl Menger, *Notices of the A.M.S.* **43(5)** (1996), 558–561.

[M] D. E. Marshall, Removable sets for bounded analytic functions, in: *"Linear and Complex Analysis Problem Book"*, (V. P. Havin, S. V. Hruscëv and N. K. Nikolskii, eds.), Lecture Notes in Math. **1043**, Springer-Verlag, 1984, pp. 485–490.

[Ma] P. Mattila, *"Geometry of sets and measures in Euclidean space"*, Cambridge Studies in Advanced Mathematics **44**, Cambridge Univ. Press, 1995.

[MMV] P. Mattila, M. Melnikov and J. Verdera, The Cauchy integral, analytic capacity, and uniform rectifiability, *Ann. of Math.* **144** (1996), 127–136.

[Me] M. Melnikov, Analytic capacity: discrete approach and curvature of measure, *Russian Acad. Sci. Sb. Math.* **186** (1995), 827–846.

[MV] M. Melnikov and J. Verdera, A geometric proof of the L^2 boundedness of the Cauchy integral on Lipschitz graphs, *Internat. Math. Res. Notices* **7** (1995), 325–331.

[MC] Y. Meyer and R. R. Coifman, *"Wavelets, Calderón-Zygmund and multilinear operators"*, Cambridge Studies in Advanced Mathematics **48**, Cambridge Univ. Press, 1997.

[Mu] T. Murai, *"A real variable method for the Cauchy transform and analytic capacity"*, Lecture Notes in Math. **1307**, Springer-Verlag, 1998.

[NTV1] F. Nazarov, S. Treil and A. Volberg, Weak type estimates and Cotlar inequalities for Calderón-Zygmund operators on non homogeneous spaces, *Internat. Math. Res. Notices* **9** (1998), 463–487.

[NTV2] F. Nazarov, S. Treil and A. Volberg, How to prove Vitushkin's conjecture by pulling ourselver up by the hair, Preprint.

[NTV3] F. Nazarov, S. Treil and A. Volberg, *Tb* Theorems for non-homogeneous spaces, Preprint.

[S] E. M. Stein, Singular Integrals: the roles of Calderón and Zygmund, *Notices of the A.M.S.* **45(9)**, 1130–1140.

[T1] X. Tolsa, L^2-boundedness of the Cauchy integral operator for continuous measures, *Duke Math. J.* **98(2)** (1999), 269–304.

[T2] X. Tolsa, A proof of the weak $(1,1)$ inequality for singular integrals with non doubling measures based on a Calderón-Zygmund decomposition, *Publ. Mat.* **45** (2001), to appear.

[V1] J. Verdera, Removability, capacity and approximation, in: *"Complex Potential Theory"*, NATO ASI Series, Kluwer Academic Publ., Dordrecht, 1994, pp. 419–473.

[V2] J. Verdera, A new elementary proof of L^2 estimates for the Cauchy Integral on Lipschitz graphs, Manuscript from a lecture given at the Conference on Geometric and Algebraic Aspects in Several Complex Variables (Cetraro, 1994).

[V3] J. Verdera, On the $T(1)$-Theorem for the Cauchy integral, *Ark. Mat.* **38** (2000), 183–199.

[Vi] A. G. Vitushkin, The analytic capacity of sets in problems of approximation theory, *Uspekhi Mat. Nauk* **22** (1967), 141–199 (English trans.); *Math. Surveys Monographs* **22** (1967), 139–200 (in Russian).

Departament de Matemàtiques, Universitat Autònoma de Barcelona, 08193 Bellaterra, Barcelona (Spain)

E-mail address: verdera@mat.uab.es